PLANNING, PROGRAMMING, BUDGETING, AND EXECUTION
IN COMPARATIVE ORGANIZATIONS

VOLUME 2

Case Studies of Selected
Allied and Partner
Nations

MEGAN McKERNAN | STEPHANIE YOUNG | ANDREW DOWSE | JAMES BLACK
DEVON HILL | BENJAMIN J. SACKS | AUSTIN WYATT | NICOLAS JOUAN
YULIYA SHOKH | JADE YEUNG | RAPHAEL S. COHEN | JOHN P. GODGES
HEIDI PETERS | LAUREN SKRABALA

Prepared for the Commission on Planning, Programming, Budgeting, and Execution Reform
Approved for public release; distribution is unlimited

For more information on this publication, visit **www.rand.org/t/RRA2195-2**.

About RAND

The RAND Corporation is a research organization that develops solutions to public policy challenges to help make communities throughout the world safer and more secure, healthier and more prosperous. RAND is nonprofit, nonpartisan, and committed to the public interest. To learn more about RAND, visit www.rand.org.

Research Integrity

Our mission to help improve policy and decisionmaking through research and analysis is enabled through our core values of quality and objectivity and our unwavering commitment to the highest level of integrity and ethical behavior. To help ensure our research and analysis are rigorous, objective, and nonpartisan, we subject our research publications to a robust and exacting quality-assurance process; avoid both the appearance and reality of financial and other conflicts of interest through staff training, project screening, and a policy of mandatory disclosure; and pursue transparency in our research engagements through our commitment to the open publication of our research findings and recommendations, disclosure of the source of funding of published research, and policies to ensure intellectual independence. For more information, visit www.rand.org/about/principles.

RAND's publications do not necessarily reflect the opinions of its research clients and sponsors.

Published by the RAND Corporation, Santa Monica, Calif.
© 2024 RAND Corporation
RAND® is a registered trademark.

Library of Congress Cataloging-in-Publication Data is available for this publication.

ISBN: 978-1-9774-1253-9

Cover design by Peter Soriano; adimas/Adobe Stock images.

About This Report

The U.S. Department of Defense (DoD) Planning, Programming, Budgeting, and Execution (PPBE) process is a key enabler for DoD to fulfill its mission. But in light of a dynamic threat environment, increasingly capable adversaries, and rapid technological changes, there has been increasing concern that DoD's resource planning processes are too slow and inflexible to meet warfighter needs.[1] As a result, Congress mandated the formation of a legislative commission in Section 1004 of the National Defense Authorization Act for Fiscal Year 2022 to (1) examine the effectiveness of the PPBE process and adjacent DoD practices, particularly with respect to defense modernization; (2) consider potential alternatives to these processes and practices to maximize DoD's ability to respond in a timely manner to current and future threats; and (3) make legislative and policy recommendations to improve such processes and practices for the purposes of fielding the operational capabilities necessary to outpace near-peer competitors, providing data and analytical insight, and supporting an integrated budget that is aligned with strategic defense objectives.[2]

The Commission on PPBE Reform requested that the National Defense Research Institute provide an independent analysis of PPBE-like functions in selected other countries and other federal agencies. This report, part of a four-volume set, analyzes the budgeting processes of allied countries and partners. Volume 1 analyzes the defense budgeting processes of China and Russia. Volume 3 analyzes the defense budgeting processes of other federal agencies. And Volume 4, an executive summary, distills key insights from these three analytical volumes. The commission will use insights from these analyses to derive potential lessons for DoD and recommendations to Congress on PPBE reform.

This report should be of interest to those concerned with the improvement of DoD's PPBE processes. The intended audience is mostly government officials responsible for such processes. The research reported here was completed in March 2023 and underwent security review with the sponsor and the Defense Office of Prepublication and Security Review before public release.

---

[1] See, for example, Section 809 Panel, *Report of the Advisory Panel on Streamlining and Codifying Acquisition Regulations*, Vol. 2 of 3, June 2018, pp. 12–13; Brendan W. McGarry, *DOD Planning, Programming, Budgeting, and Execution: Overview and Selected Issues for Congress*, Congressional Research Service, R47178, July 11, 2022, p. 1; and William Greenwalt and Dan Patt, *Competing in Time: Ensuring Capability Advantage and Mission Success Through Adaptable Resource Allocation*, Hudson Institute, February 2021, pp. 9–10.

[2] Public Law 117–81, National Defense Authorization Act for Fiscal Year 2022, December 27, 2021.

RAND National Security Research Division

This research was sponsored by the Commission on PPBE Reform and conducted within the Acquisition and Technology Policy Program of the RAND National Security Research Division (NSRD), which operates the National Defense Research Institute (NDRI), a federally funded research and development center sponsored by the Office of the Secretary of Defense, the Joint Staff, the Unified Combatant Commands, the Navy, the Marine Corps, the defense agencies, and the defense intelligence enterprise.

For more information on the RAND Acquisition and Technology Policy Program, see www.rand.org/nsrd/atp or contact the director (contact information is provided on the webpage).

Acknowledgments

The authors thank the members of the Commission on PPBE Reform—Robert Hale, Ellen Lord, Jonathan Burks, Susan Davis, Lisa Disbrow, Eric Fanning, Peter Levine, Jamie Morin, David Norquist, Diem Salmon, Jennifer Santos, Arun Seraphin, Raj Shah, and John Whitley—and staff for their dedication and deep expertise in shaping this work. We extend special gratitude to the commission chair, the Honorable Robert Hale; the vice chair, the Honorable Ellen Lord; executive director Lara Sayer; and director of research Elizabeth Bieri for their guidance and support throughout this analysis. We would also like to thank the subject-matter experts on Australia, Canada, and the United Kingdom who provided us valuable insight on these countries' PPBE-like processes.

From NSRD, we thank Barry Pavel, vice president and director, and Mike Spirtas, associate director, along with then–acting director Christopher Mouton and associate director Yun Kang of NSRD's Acquisition and Technology Policy Program, for their counsel and tireless support. We also thank our RAND Corporation colleagues who provided input at various stages of this work, including Don Snyder, Michael Kennedy, Irv Blickstein, Brian Persons, Chad Ohlandt, Bonnie Triezenberg, Obaid Younossi, Clinton Reach, John Yurchak, Jeffrey Drezner, Brady Cillo, and Gregory Graff, as well as the team of peer reviewers who offered helpful feedback on individual case studies and on cross–case study takeaways: Cynthia Cook, Roger Lough, Paul DeLuca, Hans Pung, and Jim Powers. Finally, we would like to thank Maria Falvo and Saci Haslam for their administrative assistance during this effort. The work is much improved by virtue of their inputs, but any errors remain the responsibility of the authors alone.

Dedication

These volumes are dedicated to Irv Blickstein, whose decades of experience in the U.S. Navy's PPBE community deeply informed this work and whose intellectual leadership as a RAND colleague for more than 20 years greatly enhanced the quality of our independent analysis for DoD's most-pressing acquisition challenges. Irv's kindness, motivation, and ever-present mentoring will be sorely missed.

Summary

Issue

The U.S. Department of Defense's (DoD's) Planning, Programming, Budgeting, and Execution (PPBE) System was originally developed in the 1960s as a structured approach for planning long-term resource development, assessing program cost-effectiveness, and aligning resources to strategies. Yet changes to the strategic environment, the industrial base, and the nature of military capabilities have raised the question of whether U.S. defense budgeting processes are still well aligned with national security needs.

Congress, in its National Defense Authorization Act for Fiscal Year 2022, called for the establishment of a Commission on PPBE Reform, which took shape as a legislative commission in 2022.[3] As part of its data collection efforts, the Commission on PPBE Reform asked the National Defense Research Institute, a federally funded research and development center operated by the RAND National Security Research Division, to conduct case studies of budgeting processes across nine comparative organizations: five international defense organizations and four other U.S. federal government agencies. The two international case studies of near-peer competitors China and Russia were specifically requested by Congress, while the other seven cases were selected in close partnership with the commission.

Approach

For all nine case studies, the research entailed extensive document reviews and structured discussions with subject-matter experts having experience in the budgeting processes of the selected international governments and other U.S. federal government agencies. Each case study was assigned a unique team with appropriate regional or organizational expertise. The analysis was also supplemented by experts in the U.S. PPBE process, as applicable.

Key Insights

The key insights from the case studies of selected allied and partner nations—Australia, Canada, and the United Kingdom (UK)—are as follows:

- **Australia, Canada, and the UK have a shared commitment to democratic political institutions with the United States and converge on a similar strategic vision.**

---

3 Public Law 117-81, National Defense Authorization Act for Fiscal Year 2022, December 27, 2021.

This alignment not only presents opportunities for co-development and broader prospects for working together toward shared goals but also requires the United States and its allies and partners to develop more-effective partnership approaches. In addition, each country struggles to balance the needs to keep pace with strategic threats, execute longer-term plans, use deliberate processes with sufficient oversight, and encourage innovation.

- **Foreign military sales (FMS) are important mechanisms for strategic convergence but pose myriad challenges for coordination and resource planning.** Australia, Canada, and the UK rely on U.S. FMS to promote strategic convergence, interconnectedness, interoperability, interchangeability, and the shared benefits of innovation. One downside to this reliance is that exchange-rate volatility can require unexpected budget adjustments. Another downside is that each country is less able to independently act with flexibility.
- **The Australian, Canadian, and UK political systems shape the roles and contours of resource planning.** In all three countries, the executive branch has the power of the purse, which reduces political friction over appropriations.
- **Australia, Canada, and the UK have less legislative intervention in budgeting processes, relative to the United States, and do not need to confront the challenges of operating without a regular appropriation (as is the case under continuing resolutions).** These countries' resource management systems have less partisan interference than the United States' system, according to subject-matter experts.
- **Strategic planning mechanisms in Australia, Canada, and the UK harness defense spending priorities and drive budget execution.** Each country starts its defense resource management processes with strategic planning to identify key priorities for finite funds in defense budgets that are smaller than that of the United States.
- **Jointness in resource planning appears to be easier in Australia, Canada, and the UK, given the smaller size and structure of their militaries.** In each country, there is a greater level of joint financial governance than in the United States, with less focus on service-centric views and more focus on cross-governmental mechanisms and joint funds.
- **Australia, Canada, and the UK place a greater emphasis on budget predictability and stability than on agility.** Australia's Department of Defence is assured of sustained funding for four years and plans investments as far as 20 years out. The notional budget of Canada's Department of National Defence is guaranteed to continue year on year, and the department's Capital Investment Fund ensures that approved projects are paid for years or even decades in advance. UK Ministry of Defence programs are normally guaranteed funding for three to five years, with estimates out to ten years. In contrast, Congress must revisit and vote on DoD's entire budget every year.
- **Despite the common emphasis on stability, each system provides some budget flexibility to address unanticipated changes.** The Australian Parliament can boost the defense budget in periods of national emergency or to fund overseas military opera-

tions, and the government can supplement defense allocations to alleviate inflationary pressures. In Canada, regular supplementary parliamentary spending periods can help close unforeseen defense funding gaps. The UK Ministry of Defence has mechanisms for moving money between accounts (e.g., a process known as *virement* for reallocating funds with either Treasury or parliamentary approval, depending on the circumstances) and accessing additional funds in a given fiscal year.

- **Similar budget mechanisms are used in Australia, Canada, and the UK.** All three countries carry over funds, move funds across portfolios, appropriate funds with different expirations, and supplement funds for emerging needs. The use of these mechanisms, however, varies across the cases.

- **Australia, Canada, and the UK have all pivoted toward supporting agility and innovation in the face of lengthy acquisition cycles.** The proposed Australian Strategic Capabilities Accelerator would be required to move funds between projects to accelerate innovation. Canada, whose strategic plan calls for its Department of National Defence to exploit defense innovation,[4] is partnering with the United States to modernize the North American Aerospace Defense Command (NORAD). Like DoD, the UK Ministry of Defence is experimenting with ways to encourage innovation, including a new Innovation Fund, which allows the chief scientific adviser to pursue higher-risk projects as part of the primary research and development budget.

- **Australia, Canada, and the UK have independent oversight functions for ensuring the transparency, audits, or contestability of budgeting processes.** Accountability in Australia is provided through the Australian National Audit Office, the Portfolio Budget Statement, the contestability function, and other reviews. Parliamentary oversight—or scrutiny—in Canada is aided by analyses from the Auditor General, the Parliamentary Budget Officer, and, at times, the Library of Parliament. Each year, the UK Ministry of Defence budget is externally vetted by the House of Commons Public Accounts Committee, the UK National Audit Office, and the Comptroller and Auditor General to ensure that funds are not misused.

- **Despite the push to accept additional risk, there is still a cultural aversion to risk in the Australian, Canadian, and British budgeting processes.** In Australia, stakeholders seek to spend within annual budget limits, which is intuitively prudent but could limit agility by lengthening review times and holding up funds for other projects. Canada's political structure does not allow its parliament to drastically change funding for departments, including the Department of National Defence, beyond what has been requested. The experiments by the UK Ministry of Defence to encourage innovation have not made its culture less risk-averse.

---

[4]   Canadian Department of National Defence, *Department of National Defence and Canadian Armed Forces, 2022–2023: Departmental Plan*, 2022a.

The Commission on PPBE Reform is looking for potential lessons from the PPBE-like systems of selected allied and partner nations to improve DoD's PPBE System. Of particular concern for DoD is its yearly vulnerability to political gridlock, continuing resolutions, and potential government shutdowns—all of which are obstacles that allies and partners do not endure. Without altering the U.S. system of government, which deliberately empowers strong voices from both the executive and legislative branches in defense budget decisionmaking, the United States could learn from allied and partner budgetary mechanisms that provide extra budget surety for major multiyear investments without requiring their reevaluation every year.

For example, the UK defense budgeting system benefits from multiannual spending plans, programs, and contracts. The Ministry of Defence can sign decade-long portfolio management agreements with UK firms to provide long-term certainty. The UK system also allows for advance funding early in a budget year to ensure continuous government operations, thereby avoiding the possibility—and cost—of a shutdown. Likewise, Australia's defense budgeting processes provide a high level of certainty for the development and operationalization of major military capabilities. These farsighted processes strengthen the link between strategy and resources, reduce the prospects for misused funds or inefficiency, and limit the risk of blocked funding from year to year.

Contents

Figures and Tables

Figures

Tables

CHAPTER 1

Introduction

In light of a dynamic threat environment, increasingly capable adversaries, and rapid technological changes, there has been increasing concern that the U.S. Department of Defense's (DoD's) resource planning processes are too slow and inflexible to meet warfighter needs.[1] DoD's Planning, Programming, Budgeting, and Execution (PPBE) System was originally developed in the 1960s as a structured approach for planning long-term resource development, assessing program cost-effectiveness, and aligning resources to strategies. Yet changes to the strategic environment, the industrial base, and the nature of military capabilities have raised the question of whether DoD's budgeting processes are still well aligned to national security needs.

To consider the effectiveness of current resource planning processes for meeting national security needs and to explore potential policy options to strengthen those processes, Congress called for the establishment of a commission on PPBE reform in Section 1004 of the National Defense Authorization Act for Fiscal Year (FY) 2022.[2] The Commission on PPBE Reform took shape as a legislative commission in 2022, consisting of 14 appointed commissioners, each drawing on deep and varied professional expertise in DoD, Congress, and the private sector. In support of this work, the commission collected data, conducted analyses, and developed a broad array of inputs from external organizations, including federally funded research and development centers, to develop targeted insights of particular interest to the commission. The commission asked the RAND National Defense Research Institute to contribute to this work by conducting case studies of nine comparative organizations: five international defense organizations and four other U.S. federal government agencies. Two

---

[1] See, for example, Section 809 Panel, *Report of the Advisory Panel on Streamlining and Codifying Acquisition Regulations*, Vol. 2 of 3, June 2018, pp. 12–13; Brendan W. McGarry, *DOD Planning, Programming, Budgeting, and Execution: Overview and Selected Issues for Congress*, Congressional Research Service, R47178, July 11, 2022, p. 1; and William Greenwalt and Dan Patt, *Competing in Time: Ensuring Capability Advantage and Mission Success Through Adaptable Resource Allocation*, Hudson Institute, February 2021, pp. 9–10.

[2] Public Law 117-81, National Defense Authorization Act for Fiscal Year 2022, December 27, 2021. Section 1004(f) of this Act is of particular relevance to our research approach:

> Compare the planning, programming, budgeting, and execution process of the Department of Defense, including the development and production of documents including the Defense Planning Guidance (described in section 113(g) of Title 10, United States Code), the Program Objective Memorandum, and the Budget Estimate Submission, with similar processes of private industry, other Federal agencies, and other countries.

of the international case studies—of near-peer competitors—were specifically called for by Congress, and additional cases were selected in close partnership with the commission.[3]

This report is Volume 2 in a four-volume set, three of which present case studies conducted in support of the Commission on PPBE Reform. The accompanying volumes focus on selected near-peer competitors (*Planning, Programming, Budgeting and Execution in Comparative Organizations*: Vol. 1, *Case Studies of China and Russia*) and selected U.S. federal government agencies (*Planning, Programming, Budgeting and Execution in Comparative Organizations*: Vol. 3, *Case Studies of Selected Non-DoD Federal Agencies*).[4] Volume 4, an executive summary, distills key insights from these three analytical volumes.[5]

Evolution of DoD's PPBE System

The Planning, Programming, and Budgeting System (PPBS), the precursor to DoD's PPBE process, took shape in the first decades after World War II and was introduced into DoD in 1961 by then–Secretary of Defense Robert McNamara.[6] Drawing on new social science methods, such as program budgeting and systems analysis, the PPBS was designed to provide a structured approach to weigh the cost-effectiveness of potential defense investments. A central assertion of the PPBS's developers was that strategy and costs needed to be considered together.[7] As Charles Hitch, Secretary McNamara's first comptroller and a key intellectual leader in the development and implementation of the PPBS, noted, "There is no budget size

---

[3] Pub. L. 117-81, Section 1004(f) requires "a review of budgeting methodologies and strategies of near-peer competitors to understand if and how such competitors can address current and future threats more or less successfully than the United States."

[4] Megan McKernan, Stephanie Young, Timothy R. Heath, Dara Massicot, Mark Stalczynski, Ivana Ke, Raphael S. Cohen, John P. Godges, Heidi Peters, and Lauren Skrabala, *Planning, Programming, Budgeting, and Execution in Comparative Organizations*: Vol. 1, *Case Studies of China and Russia*, RAND Corporation, RR-A2195-1, 2024; Megan McKernan, Stephanie Young, Ryan Consaul, Michael Simpson, Sarah W. Denton, Anthony Vassalo, William Shelton, Devon Hill, Raphael S. Cohen, John P. Godges, Heidi Peters, and Lauren Skrabala, *Planning, Programming, Budgeting, and Execution in Comparative Organizations*: Vol. 3, *Case Studies of Selected Non-DoD Federal Agencies*, RAND Corporation, RR-A2195-3, 2024.

[5] Megan McKernan, Stephanie Young, Timothy R. Heath, Dara Massicot, Andrew Dowse, Devon Hill, James Black, Ryan Consaul, Michael Simpson, Sarah W. Denton, Anthony Vassalo, Ivana Ke, Mark Stalczynski, Benjamin J. Sacks, Austin Wyatt, Jade Yeung, Nicolas Jouan, Yuliya Shokh, William Shelton, Raphael S. Cohen, John P. Godges, Heidi Peters, and Lauren Skrabala, *Planning, Programming, Budgeting, and Execution in Comparative Organizations*: Vol. 4, *Executive Summary*, RAND Corporation, RR-A2195-4, 2024.

[6] An oft-quoted assertion by Secretary McNamara from April 20, 1963, which is pertinent to this discussion, is that "[y]ou cannot make decisions simply by asking yourself whether something might be nice to have. You have to make a judgment on how much is enough" (as cited in the introduction of Alain C. Enthoven and K. Wayne Smith, *How Much Is Enough? Shaping the Defense Program, 1961–1969*, RAND Corporation, CB-403, 1971).

[7] Or, as Bernard Brodie stated succinctly, "strategy wears a dollar sign" (Bernard Brodie, *Strategy in the Missile Age*, RAND Corporation, CB-137-1, 1959, p. 358).

or cost that is correct regardless of the payoff, and there is no need that should be met regardless of cost."[8]

To make decisions about prioritization and where to take risk in a resource-constrained environment, DoD needed an analytic basis for making choices. Therefore, the PPBS first introduced the program budget, an *output*-oriented articulation of the resources associated with a given military capability projected out over five years.[9] Second, the PPBS introduced an approach for assessing cost-effectiveness, termed *systems analysis*, which was institutionalized in an Office of Systems Analysis. Since 2009, this office has been known as Cost Assessment and Program Evaluation (CAPE).[10] At its inception, the PPBS was a process for explicitly linking resources to strategy and for setting up a structure for making explicit choices between options, based on the transparent analysis of costs and effectiveness. Then, as today, the system introduced friction with other key stakeholders, including Congress and industry partners. Key features of the PPBS have become institutionalized in DoD's PPBE System, and questions have arisen about whether its processes and structures remain relevant and agile enough to serve their intended purposes.[11]

To set up the discussion of case studies, it will be helpful to outline the key features of the PPBE process and clarify some definitions. Figure 1.1 offers a summary view of the process.

---

[8] Charles J. Hitch and Roland N. McKean, *The Economics of Defense in the Nuclear Age*, RAND Corporation, R-346, 1960, p. 47.

[9] On the need for an output-oriented budget formulation at the appropriate level to make informed choices, Hitch and McKean (1960, p. 50) noted that the consumer "cannot judge intelligently how much he should spend on a car if he asks, 'How much should I devote to fenders, to steering activities, and to carburetion?' Nor can he improve his decisions much by lumping all living into a single program and asking, 'How much should I spend on life?'"

[10] In an essential treatise on the PPBS's founding, Enthoven (the first director of the Office of Systems Analysis) and Smith describe "the basic ideas that served as the intellectual foundation for PPBS" (1971, pp. 33–47) and, thus, PPBE: (1) decisionmaking should be made on explicit criteria of the national interest, (2) needs and costs should be considered together, (3) alternatives should be explicitly considered, (4) an active analytic staff should be used, (5) a multiyear force and financial plan should project consequences into the future, and (6) open and explicit analysis should form the basis for major decisions.

[11] Greenwalt and Patt, 2021, pp. 9–10.

FIGURE 1.1

DoD's PPBE Process (as of September 2019)

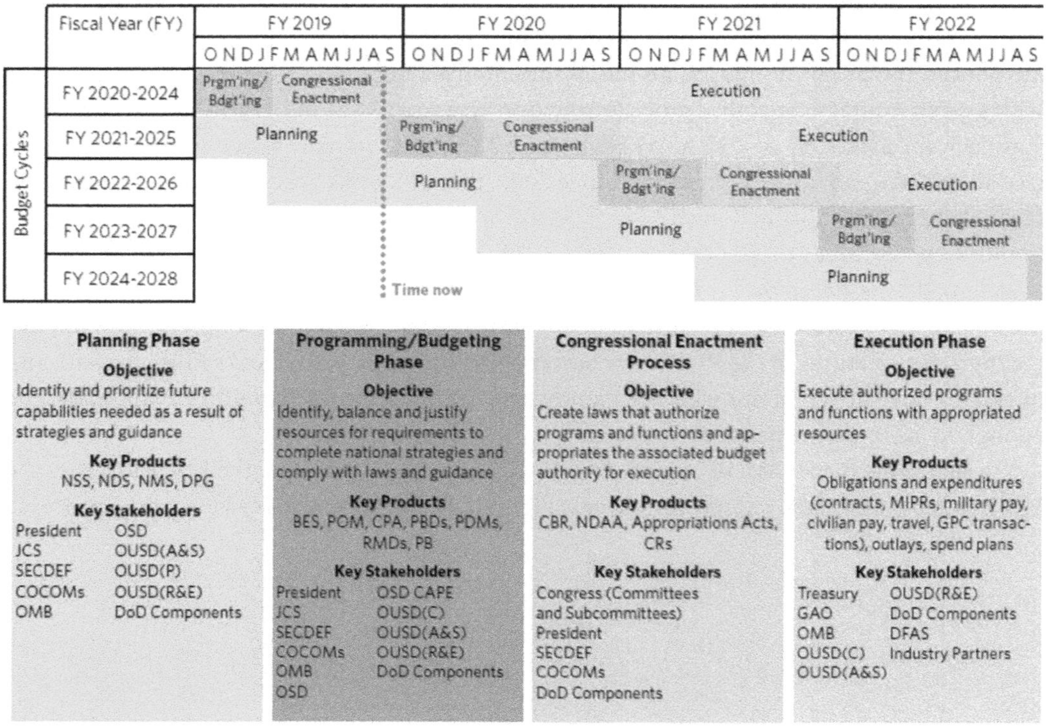

Fiscal Year (FY)	FY 2019	FY 2020	FY 2021	FY 2022
	O N D J F M A M J J A S	O N D J F M A M J J A S	O N D J F M A M J J A S	O N D J F M A M J J A S
FY 2020-2024	Prgm'ing/ Bdgt'ing · Congressional Enactment	Execution		
FY 2021-2025	Planning	Prgm'ing/ Bdgt'ing · Congressional Enactment	Execution	
FY 2022-2026		Planning	Prgm'ing/ Bdgt'ing · Congressional Enactment	Execution
FY 2023-2027			Planning	Prgm'ing/ Bdgt'ing · Congressional Enactment
FY 2024-2028	Time now			Planning

Budget Cycles

Planning Phase	Programming/Budgeting Phase	Congressional Enactment Process	Execution Phase
Objective Identify and prioritize future capabilities needed as a result of strategies and guidance	**Objective** Identify, balance and justify resources for requirements to complete national strategies and comply with laws and guidance	**Objective** Create laws that authorize programs and functions and appropriates the associated budget authority for execution	**Objective** Execute authorized programs and functions with appropriated resources
Key Products NSS, NDS, NMS, DPG	**Key Products** BES, POM, CPA, PBDs, PDMs, RMDs, PB	**Key Products** CBR, NDAA, Appropriations Acts, CRs	**Key Products** Obligations and expenditures (contracts, MIPRs, military pay, civilian pay, travel, GPC transactions), outlays, spend plans
Key Stakeholders President · OSD JCS · OUSD(A&S) SECDEF · OUSD(P) COCOMs · OUSD(R&E) OMB · DoD Components	**Key Stakeholders** President · OSD CAPE JCS · OUSD(C) SECDEF · OUSD(A&S) COCOMs · OUSD(R&E) OMB · DoD Components OSD	**Key Stakeholders** Congress (Committees and Subcommittees) President SECDEF COCOMs DoD Components	**Key Stakeholders** Treasury · OUSD(R&E) GAO · DoD Components OMB · DFAS OUSD(C) · Industry Partners OUSD(A&S)

SOURCE: Reproduced from Stephen Speciale and Wayne B. Sullivan II, "DoD Financial Management—More Money, More Problems," Defense Acquisition University, September 1, 2019, p. 6.

NOTE: BES = budget estimation submission; CBR = concurrent budget resolution; COCOM = combatant command; CPA = Chairperson's Program Assessment; CR = continuing resolution; DFAS = Defense Finance and Accounting Services; DPG = defense planning guidance; GAO = U.S. Government Accountability Office; GPC = government purchase card; JCS = Joint Chiefs of Staff; MIPR = military interdepartmental purchase request; NDAA = National Defense Authorization Act; NDS = National Defense Strategy; NMS = National Military Strategy; NSS = National Security Strategy; OMB = Office of Management and Budget; OSD = Office of the Secretary of Defense; OUSD(A&S) = Office of the Under Secretary of Defense (Acquisition and Sustainment); OUSD(C) = Office of the Under Secretary of Defense (Comptroller); OUSD(P) = Office of the Under Secretary of Defense (Policy); OUSD(R&E) = Office of the Under Secretary of Defense (Research and Engineering); PB = President's Budget; PBD = program budget decision; PDM = program decision memorandum; POM = program objective memorandum; RMD = resource management decision; SECDEF = Secretary of Defense.

Today, consideration of PPBE often broadly encapsulates internal DoD processes, other executive branch functions, and congressional rules governing appropriations. Internal to DoD, PPBE is an annual process by which the department determines how to align strategic guidance to military programs and resources. The process supports the development of DoD inputs to the President's Budget and to a budgeting program with a five-year time horizon, known as the Future Years Defense Program (FYDP).[12] DoD Directive (DoDD) 7045.14, *The Planning, Programming, Budgeting, and Execution (PPBE) Process*, states that one intent for PPBE "is to provide the DOD with the most effective mix of forces, equipment, manpower, and support attainable within fiscal constraints."[13] PPBE consists of four distinct processes, each with its own outputs and stakeholders. Select objectives of each phase include the following:

- **Planning:** "Integrate assessments of potential military threats facing the country, overall national strategy and defense policy, ongoing defense plans and programs, and projected financial resources into an overall statement of policy."[14]
- **Programming:** "[A]nalyze the anticipated effects of present-day decisions on the future force" and detail the specific forces and programs proposed over the FYDP period to meet the military requirements identified in the plans and within the financial limits.[15]
- **Budgeting:** "[E]nsure appropriate funding and fiscal controls, phasing of the efforts over the funding period, and feasibility of execution within the budget year"; restructure budget categories for submission to Congress according to the appropriation accounts; and prepare justification material for submission to Congress.[16]
- **Execution:** "[D]etermine how well programs and financing have met joint warfighting needs."[17]

Several features of congressional appropriations processes are particularly important to note. First, since FY 1960, Congress has provided budget authority to DoD through specific appropriations titles (sometimes termed *colors of money*), the largest of which are operation and maintenance (O&M); military personnel; research, development, test, and evaluation (RDT&E); and procurement.[18] These appropriations titles are further broken down into *appropriation accounts*, such as Military Personnel, Army or Shipbuilding and Conversion, Navy (SCN). Second, the budget authority provided in one of these accounts is generally available for obligation only within a specified period. In the DoD budget, the period of availability for military personnel and O&M accounts is one year; for RDT&E accounts, two years; and for most procurement accounts, three years (although for SCN, it can be five or six years,

---

[12] Brendan W. McGarry, *Defense Primer: Planning, Programming, Budgeting and Execution (PPBE) Process*, Congressional Research Service, IF10429, January 27, 2020, p. 1.

[13] DoDD 7045.14, *The Planning, Programming, Budgeting, and Execution (PPBE) Process*, U.S. Department of Defense, August 29, 2017, p. 2.

[14] Congressional Research Service, *A Defense Budget Primer*, RL30002, December 9, 1998, p. 27.

[15] Congressional Research Service, 1998, p. 27; McGarry, 2020, p. 2.

[16] McGarry, 2020, p. 2; Congressional Research Service, 1998, p. 28.

[17] DoDD 7045.14, 2017, p. 11.

[18] Congressional Research Service, 1998, pp. 15–17.

in certain circumstances). This specification means that budget authority must be obligated within those periods or, with only a few exceptions, it is lost.[19] There has been recent interest in exploring how these features of the appropriations process affect transparency and oversight, institutional incentives, and the exercise of flexibility, should resource needs change.[20]

Importantly, PPBE touches almost everything DoD does and, thus, forms a critical touchpoint for engagement with stakeholders across DoD (e.g., OSD, military departments, Joint Staff, COCOMs), in the executive branch (through OMB), in Congress, and among industry partners.

Research Approach and Methods

In close partnership with the commission, we selected nine case studies to explore decision-making in organizations facing challenges similar to those experienced in DoD: exercising agility in the face of changing needs and enabling innovation. Two near-peer case studies were specifically called for in the legislation, in part to allow the commission to explore the competitiveness implications of strategic adversaries' approaches to resource planning.

For all nine case studies, we conducted extensive document reviews and structured discussions with subject-matter experts having experience in the budgeting processes of the international governments and other U.S. federal government agencies. For case studies of two allied and partner countries, the team leveraged researchers in RAND Europe (located in Cambridge, United Kingdom) and RAND Australia (located in Canberra, Australia) with direct experience in partner defense organizations. Given the diversity in subject-matter expertise required across the case studies, each one was assigned a unique team with appropriate regional or organizational expertise. For the near-peer competitor cases, the assigned experts had the language skills and methodological training to facilitate working with primary sources in Chinese or Russian. The analysis was also supplemented by experts in PPBE as applicable.

Case study research drew primarily on government documentation outlining processes and policies, planning guidance, budget documentation, and published academic and policy research. Although participants in structured discussions varied in accordance with the decisionmaking structures across case studies, they generally included chief financial officers, representatives from organizations responsible for making programmatic choices, and budget officials. For obvious reasons, the China and Russia case studies faced unique challenges in data collection and in identifying and accessing interview targets with direct knowledge of PPBE-like processes.

---

[19] Congressional Research Service, 1998, pp. 49–50. Regarding RDT&E, see U.S. Code, Title 10, Section 3131, Availability of Appropriations.

[20] McGarry, 2022.

To facilitate consistency, completeness in addressing the commission's highest-priority areas of interest, and cross-case comparisons, the team developed a common case study template. This template took specific questions from the commission as several inputs, aligned key questions to PPBE processes and oversight mechanisms, evaluated perceived strengths and challenges of each organization's processes and their applicability to DoD processes, and concluded with lessons learned from each case. To enable the development of a more consistent evidentiary base across cases, the team also developed a standard interview protocol to guide the structured discussions.

Areas of Focus

Given the complexity of PPBE and its many connections to other processes and stakeholders, along with other inputs and ongoing analysis by the commission, we needed to scope this work in accordance with three of the commission's top priorities.

First, although we sought insights across PPBE phases in each case study, in accordance with the commission's guidance, we placed a particular emphasis on an organization's budgeting and execution mechanisms, such as the existence of appropriations titles (i.e., colors of money), and on any mechanisms for exercising flexibility, such as reprogramming thresholds. However, it is important to note that this level of detailed information was not uniformly available. The opacity of internal processes in China and Russia made the budget mechanisms much more difficult to discern in those cases in particular.

Second, while the overall investment portfolios varied in accordance with varying mission needs, the case studies were particularly focused on investments related to RDT&E and procurement rather than O&M or sustainment activities.

Third, the case studies of other U.S. federal government agencies did not focus primarily on the roles played by external stakeholders, such as OMB, Congress, and industry partners. Such stakeholders were discussed when relevant insights emerged from other sources, but interviews and data collection were focused within the bounds of a given organization rather than across a broader network of key stakeholders.

Research Limitations and Caveats

This research required detailed analysis of the nuances of internal resource planning processes across nine extraordinarily diverse organizations and on a tight timeline required by the commission's challenging mandate. This breadth of scope was intended to provide the commission with diverse insights into how other organizations address similar challenges but also limited the depth the team could pursue for any one case. These constraints warrant additional discussion of research limitations and caveats of two types.

First, each case study, to a varying degree, confronted limitations in data availability. The teams gathered documentation from publicly available sources and doggedly pursued additional documentation from targeted interviews and other experts with direct experience, but even for the cases from allied countries and U.S. federal agencies, including DoD, there was a limit to what could be established in formal documentation. Some important features of how systems work in practice are not captured in formal documentation, and such features had to be teased out and triangulated from interviews to the extent that appropriate officials were available to engage with the team. The general opacity and lack of institutional connections to decisionmakers in China and Russia introduced unique challenges for data collection. Russia was further obscured by the war in Ukraine during the research period, which made access by U.S.-based researchers to reliable government data on current plans and resource allocation impossible.

Second, the case study teams confronted important inconsistencies across cases, which made cross-case comparability very challenging to establish. For example, international cases each involved unique political cultures, governance structures, strategic concerns, and military commitments—all of which we characterize to the extent that it is essential context for understanding how and why resource allocation decisions are made. The context-dependent nature of the international cases made even defining the "defense budget" difficult, given countries' various definitions and inclusions. With respect to the near-peer case studies of China and Russia presented in the first volume, inconsistencies were especially pronounced regarding the purchasing power within those two countries. To address some of these inconsistencies, we referenced the widely cited Stockholm International Peace Research Institute (SIPRI) Military Expenditure Database.[21] With respect to the other U.S. federal agencies, each agency had its own unique mission, organizational culture, resource level, and process of congressional oversight—all of which were critical for understanding how and why resource allocation decisions were made. This diversity strained our efforts to draw cross-case comparisons or to develop internally consistent normative judgments of best practices. For this reason, each case study analysis and articulation of strengths and challenges should be understood relative to each organization's *own* unique resource allocation needs and missions.

Selected Allied and Partner-Nations Focus

The 2022 National Defense Strategy (NDS) describes a security environment of complex strategic challenges associated with such dynamics as emerging technology, transboundary threats, and competitors posing "new threats to the U.S. homeland and to strategic stability."[22] Among these challenges, the NDS notes that "[t]he most comprehensive and serious challenge" is the People's Republic of China (PRC). The NDS points to China's military modernization and exercise of whole-of-government levers to effect "coercive" and "aggressive" approaches to the region and international order.[23] While the NDS designates China

---

[21] SIPRI, "SIPRI Military Expenditure Database," homepage, undated.

[22] DoD, *2022 National Defense Strategy of the United States of America*, 2022, p. 4.

[23] DoD, 2022, p. 4.

as the "pacing challenge" for DoD, it also highlights the threat posed by Russia as an "acute threat."[24]

To counter these strategic challenges, the NDS calls for strong relationships among U.S. allies and partner nations:

> The 2022 NDS advances a strategy focused on the PRC and on collaboration with our growing network of Allies and partners on common objectives. . . . The Department will support robust deterrence of Russian aggression against vital U.S. national interests, including our treaty Allies. . . . The 2022 National Defense Strategy is a call to action for the defense enterprise to incorporate Allies and partners at every stage of defense planning. . . . To succeed in these objectives, the Department will reduce institutional barriers, including those that inhibit collective research and development, planning, interoperability, intelligence and information sharing, and export of key capabilities.[25]

To better understand and operate in the competitive environment, the Commission on PPBE Reform is considering "budgeting methodologies and strategies of near-peer competitors to understand if and how such competitors can address current and future threats more or less successfully than the United States," along with defense resource planning in allied and partner nations.[26] For the allied and partner nations, the commission asked us to analyze the defense resource planning processes of Australia, Canada, and the United Kingdom (UK). Notably, this focus on internal processes as key enablers of military outcomes is well aligned with the NDS's imperatives of "build[ing] enduring advantage," "undertaking reforms to accelerate force development, getting the technology we need more quickly, and making investments in the extraordinary people of the Department, who remain our most valuable resource."[27] These imperatives have prompted reflection on the extent to which internal DoD processes, including PPBE, are up to the challenge of enabling rapid and responsive capability development to address the emerging threats.

The lower half of Figure 1.2 illustrates the increasing gap between China's rising military expenditure over time and the relatively flat, lower levels of expenditure by Russia and U.S. allies and partner nations: Australia, Canada, and the UK. The following sections will summarize the defense budgeting processes of those allies and partners.

Australia

Australia has a mixed system of government that includes a representative democracy, a constitutional monarchy, and a federation of states. Within this system, the Australian Depart-

---

[24] DoD, 2022, pp. 4–5.

[25] DoD, 2022, pp. 2, 14.

[26] Pub. L. 117-81, Section 1004(f)(2)(F).

[27] DoD, 2022, p. iv.

FIGURE 1.2

Military Expenditure, by Country

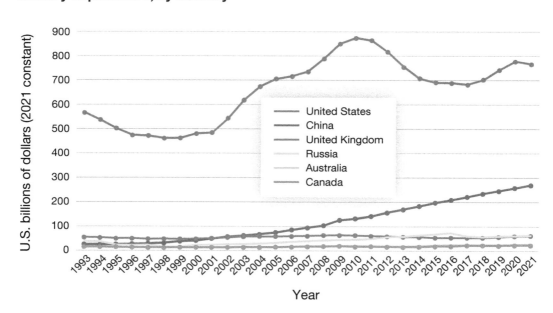

SOURCE: Features information from SIPRI, undated. Data shown are as of March 17, 2023.

ment of Defence (hereafter referred to as *Defence*) is responsible for defending Australia and its overseas territories, as well as executing defense-related missions wherever and whenever required by the national government. The Australian government aligns resources to fill Defence mission needs through its Integrated Investment Program (IIP), which is the plan for future capability investment, and its Portfolio Budget Statement, which is the proposed allocation of resources to outcomes.[28] In recent decades, Defence has participated in operations in Iraq and Afghanistan, various peacekeeping missions (e.g., in Timor-Leste), and humanitarian assistance and disaster response missions in the Indo-Pacific theater.

Defence operates in close concert with several allies, especially the United States, and leverages those alliances and partnerships as a central tool of national security. Australia is a member of the Five Eyes (Australia, Canada, New Zealand, UK, and United States) security agreement and the Quadrilateral Security Dialogue (the Quad). Australia joined the National Technology and Industrial Base (NTIB) in 2017, although the potential of allied cooperation under the NTIB has not yet been realized because of export control barriers.[29]

---

[28] The role of Portfolio Budget Statements is to inform senators and members of parliament (MPs) of the proposed allocation of resources to outcomes by each department. These documents also explain and justify those allocations with regard to outcomes in the context of the whole-of-government budget.

[29] Brendan Thomas-Noone, *Ebbing Opportunity: Australia and the US National Technology and Industrial Base*, United States Studies Centre, November 2019.

Australia is also a member of the recently signed Australia–United Kingdom–United States (AUKUS) agreement, which is designed to strengthen the defense relationship among the three countries through technological cooperation and the co-development of advanced capabilities. The strong allied focus of Australia's defense strategy emphasizes the importance of interoperability and, in some cases, integration—a key consideration in acquisition and force generation. As we discuss later, this emphasis poses a challenge to Australia's ability to independently pursue flexibility. Finally, Australia is a strategically located partner in the Indo-Pacific theater and shares U.S. concerns about China's military rise.

Defence operated with a nominal budget of Australian dollar (AUD) 48.7 billion (U.S. $34.52 billion) in FY 2022–2023.[30] This budget included funding for the Australian Signals Directorate, a distinct organization within Defence, as well as smaller defense-related agencies and programs.[31] The Australian defense budget equates to more than 2 percent of Australia's gross domestic product (GDP). This percentage reflects a continuing commitment to increasing defense spending in response to a worsening geostrategic threat environment.

Australia has a highly organized defense budgeting system; its budgeting process is based on a systemic strategy-to-task approach, in which there is clear alignment between resources and the outcomes that they deliver. Australia's budget allocations are expected to rise as it prepares to develop its military to respond to new and emerging threats in the Indo-Pacific.

A key strength of the Australian system is that Defence can lay out its baseline budget over a ten-year period through strategic documents, most recently the *2016 Defence White Paper*.[32] Defence, therefore, enjoys an important degree of budget surety. Another strength is that the unapproved, fungible portions of the IIP (which is reviewed biannually) provide fiscal flexibility for Australia's military services and programs.[33] Funding from unapproved projects in the IIP can be shifted to introduce, cancel, prioritize, or deprioritize capability programs biannually in response to external factors. The smaller and more integrated nature of the Australian Defence Force, relative to that of the U.S. military services, allows for funding to be shifted across the services as needs arise. The IIP provides public transparency and prepares industry through demand signals of potential future requirements.[34]

A key challenge is that the Australian One Defence Capability System (ODCS), like DoD's PPBE System, can take many years (up to a decade) to allow a capability program to pass

---

[30] Australian Department of Defence, "Budgets," webpage, undated-a.

[31] For a more in-depth look at the Australian defense organizational structure, see Australian Department of Defence, *Defence Corporate Plan: 2022–2026*, August 2022a, p. 17; and Australian Department of Defence, *Portfolio Budget Statements 2022–23: Defence Portfolio—Entity Resources and Planned Performance*, October 2022c, p. 10, Figure 2.

[32] Australian Defence official, interview with the authors, October 2022. See Australian Department of Defence, *2016 Defence White Paper*, 2016b.

[33] Australian Defence official, interview with the authors, October 2022. The three Australian military services are the Royal Australian Navy, the Australian Army, and the Royal Australian Air Force.

[34] David Watt and Nicole Brangwin, "Defence Capability," Australian Parliament, webpage, undated.

through the ODCS and the IIP from unfunded concept to funded reality. Finally, given the close defense cooperation relationship between Australia and the United States, any changes to DoD's PPBE System likely will lead to secondary effects for the Australian system.

Canada

Canada has a mixed system of government: It is a federation of provinces and territories, a constitutional monarchy, and a parliamentary democracy. It has a strong central federal government led by the parliament, which shares domestic policy responsibilities with the governments of the country's ten provinces and three territories.

Canada and the United States have a long, collaborative defense relationship. Their militaries have fought alongside one another in several conflicts since World War II. Canada describes the United States as its "most important ally and defence partner,"[35] and the U.S. Department of State says that the two countries' "bilateral relationship is one of the closest and most extensive."[36] Both Canada and the United States are members of the North Atlantic Treaty Organization (NATO), and they cooperate extensively through multiple bilateral defense forums and agreements, including the North American Aerospace Defense Command (NORAD), the Permanent Joint Board on Defense, the Military Cooperation Committee, the Combined Defense Plan, the Tri-Command Framework, and the Canada-U.S. Civil Assistance Plan.[37]

Despite this close relationship, the United States and Canada spend vastly different amounts on defense annually: The United States appropriated about U.S. $798 billion in FY 2023, while Canada spent roughly one-40th of that sum, or U.S. $19 billion, in FY 2022–2023.[38] Canada's parliamentary system and the U.S. political system also operate very differently. Nonetheless, a review of Canada's defense budgeting process can provide U.S. policymakers with useful insights on resource allocation methods and challenges.

For example, a key strength is that Canada recognizes its status as a middle power and has sought to increase its relative influence through multilateral diplomacy and contributions to alliances;[39] this alliance-oriented foreign and defense policy approach has helped offset relative personnel and resource limitations. Furthermore, the Canadian Department of

---

[35] Government of Canada, "The Canada-U.S. Defence Relationship," webpage, updated April 25, 2014.

[36] U.S. Department of State, "U.S. Relations with Canada," fact sheet, August 19, 2022.

[37] Government of Canada, 2014.

[38] U.S. House of Representatives, Committee on Appropriations, *Defense: Fiscal Year 2023 Appropriations Bill Summary*, U.S. Government Publishing Office, undated; Government of Canada, "Main Estimates—2022–23 Estimates," webpage, updated March 1, 2022b. The Canadian government's fiscal year runs from April 1 to March 30 of the following year. For example, FY 2022–2023 began on April 1, 2022, and ended on March 30, 2023.

[39] Peter J. Meyer and Ian F. Fergusson, *Canada-U.S. Relations*, Congressional Research Service, No. 96-397, February 10, 2021.

National Defence (DND) has sought to improve budget transparency in recent years, allowing for better long-term spending projections, though challenges remain that we discuss later. The DND has also taken on a service-agnostic acquisition process that weighs new projects against strategic priorities.

In terms of challenges, there appears to be little political appetite for defense spending growth in Canada, which limits Canada's ability to quickly reach NATO's goal of spending 2 percent of GDP on defense. There is also limited bureaucratic capacity to absorb that new spending quickly.

United Kingdom

The UK is a constitutional monarchy with a bicameral parliamentary system. The stability of the bicameral system relies on the fact that the chief of the executive branch (the prime minister, formally the First Lord of the Treasury) is a member of parliament from whichever party is able to command the confidence of a majority of the elected lower chamber, the House of Commons.[40] The upper chamber, the House of Lords, is not elected but appointed. The government of the day may or may not hold a majority in the House of Lords, whose function is largely to offer advice and scrutiny. By centuries-old convention, the upper chamber defers to the lower chamber on financial matters, limiting the upper chamber's ability to amend or block spending bills.

This interweaving of the executive and legislature, along with the use of a "first-past-the-post" (or plurality) voting system to elect members of parliament, is intended to empower the prime minister and the prime minister's chosen cabinet to rule with a strong mandate. Because the UK's government necessarily emerges from the parliament's majority, there is less inherent antagonism between the branches of government than in the United States. The resulting empowerment of the prime minister can enable more-streamlined executive and legislative action, but it also limits the formal checks and balances that characterize the U.S. system.

Within this structure, parliament must approve the government's defense missions and the resources that the UK Ministry of Defence (MoD) requests to perform those missions. Without this approval, there are consequences for the prime minister: de facto opposition from the prime minister's own majority in the House of Commons triggering a no-confidence vote and the likely collapse of the current government. The alignment of resource allocation with the MoD's mission is therefore a structural feature of the UK parliamentary system, at least as far as the government properly estimates the resources needed to satisfy its defense needs.

---

[40] This can be via a party winning an outright majority of seats (as is typically the case), by entering a formal coalition with one or more other parties (as the Conservatives and Liberal Democrats did in 2010–2015), or through a looser arrangement known as "confidence and supply," whereby a minority party rules alone but with other parties agreeing to back it on votes of confidence or supply even if they do not enter into a formal coalition government.

The UK government is committed to maintaining defense spending above 2 percent of GDP, in line with NATO targets. Given the MoD's ambitious long-term goals and concurrent requirement to respond to near-term operational pressures, it will need to overcome both internal barriers (e.g., bureaucracy) and the destabilizing impact of a confluence of the following external trends:

- The UK has experienced an unprecedented period of acute political instability (e.g., three prime ministers in 2022) and faces increased fiscal pressures in the wake of Brexit, the coronavirus disease 2019 (COVID-19) pandemic, and cost-of-living and energy crises.
- The war in Ukraine has tested the flexibility of the UK's budgetary mechanisms in responding to emerging and unplanned requirements. Aid packages to Ukraine have depleted defense equipment and munitions stockpiles; the UK's support for Ukraine is second only to that of the United States.
- Inflation has been increasing sharply and might force the MoD to cut its budget in real terms. Similarly, the defense sector is highly exposed to foreign exchange rate (trading) volatility, given the extent of its U.S. imports, which are primarily aircraft (e.g., F-35Bs, P-8s, AH-64 *Apache* helicopters, CH-47 *Chinook* helicopters).
- The UK's expeditionary focus, international presence through its Overseas Territories, and global commitments (which are similar to those of the United States) require a broad mix of capabilities and equipment for diverse conditions and terrain. This requirement sets the UK apart from other medium powers, such as France, Germany, and Japan, which have narrower mission sets.
- Cost growth and escalation challenges have been intensified by the industrial base and supply chain challenges over the past few years.
- The MoD is under increased pressure to use its budget to boost economic prosperity (through jobs, exports, and so on) and maximize the environmental and social benefits of spending. New Treasury rules on public procurement give a minimum 10-percent weighting to social value in contract award decisions.[41]

This mix of long-term and immediate pressures poses significant dilemmas for UK defense planners and those responsible for managing the MoD's finances and executing its spending plans. But the pressures also provide an impetus for ongoing efforts to adapt the

---

[41] The 2021 *Defence and Security Industrial Strategy* introduced a requirement for all defense contract awards to consider the broader social value of spending, with a minimum of 10-percent weighting in the overall evaluation criteria (UK Ministry of Defence, *Defence and Security Industrial Strategy: A Strategic Approach to the UK's Defence and Security Industrial Sectors*, March 2021b). This requirement reflects a wider shift in HM Treasury *Green Book* guidance on appraisals and evaluation for public-sector contracts (HM Treasury, "The Green Book," webpage, updated November 18, 2022b). *Social value* in this context can include "economic (e.g. employment or apprenticeship/training opportunities), social (e.g. activities that promote cohesive communities) and environmental (e.g. efforts in reducing carbon emissions)" benefits (UK Government Commercial Function, *Guide to Using the Social Value Model*, edition 1.1, December 3, 2020, pp. 2–3).

MoD's PPBE processes to encourage more agility and innovation, improve value for money across the portfolio, and enable the MoD and the military to deliver increased output despite limited resources.

Given the UK's significance as a defense actor, DoD could draw lessons from its past and ongoing efforts to promote flexibility, agility, and innovation. Moreover, it is important for U.S. defense leaders to understand the MoD's budgeting cycle because the UK is a critical ally that retains global military responsibilities and capabilities, including nuclear weapons. The UK is a member of the trilateral AUKUS security pact, the Combined Joint Expeditionary Force with France, the European Intervention Initiative, the Five Eyes security agreement, the Five Power Defence Arrangements (with Australia, Malaysia, New Zealand, and Singapore), the Joint Expeditionary Force (which the UK leads), NATO, and the Northern Group. The UK is also one of five veto-wielding permanent members of the United Nations Security Council. Therefore, the MoD interacts frequently and interoperates closely with the U.S. intelligence community and military, and its defense budget and planning decisions are often made in unofficial concert with DoD decisions and priorities.

Although the MoD operates in a different constitutional, political, and fiscal context from its U.S. counterpart, its approach to PPBE could hold insights for DoD. For example, although UK government departments are subject to three- to five-year spending reviews, they are not subject to legislative interference or continuing resolutions. This approach gives defense planners a valuable degree of certainty.[42] Multiyear spending reviews make budgeting more rigid than a yearly budget would, but the Treasury and MoD retain some flexibility when translating the medium-term vision of comprehensive spending reviews into annual budgets and plans. For instance, the UK has mechanisms—although imperfect and perhaps not always used as widely as needed—for moving money between accounts and for accessing additional funds in a given fiscal year. These mechanisms include a process known as *virement* for reallocating funds with either Treasury or parliamentary approval, depending on the circumstances. The MoD can make additional funding requests through in-year supplementary estimates sent to parliament. The MoD has other types of flexibility as well, including access to additional Treasury funds to cover urgent capability requirements (UCRs), and it can use the cross-governmental UK Integrated Security Fund (UKISF[43])—formerly, the Conflict, Stability and Security Fund (CSSF)—or the Deployed Military Activity Pool "to

---

[42] HM Treasury (also known as *the Exchequer*) acts as both the treasury and finance ministry and owns the Public Spending Framework. It has a statutory responsibility for setting departmental budgets across the government and is internally political to the governing party but not the parliament, ensuring a degree of stability to implement long-term policies (UK subject-matter expert, interview with the authors, October 2022).

[43] UK Cabinet Office, "PM Announces Major Defence Investment in Launch of Integrated Review Refresh," press release, March 13, 2023.

make available resources to fund the initial and short-term costs of unforeseen military activity," such as responses to natural disasters or support to Ukraine.[44]

Like DoD, the MoD is experimenting with new ways to encourage innovation, including a new dedicated Innovation Fund, which allows the chief scientific adviser to pursue higher-risk projects as part of the main research and development (R&D) budget. The MoD has further supported innovation through incubators, accelerators, and novel contracting practices.

However, these strategies have not alleviated some enduring challenges in the MoD budget process. The challenges include a risk-averse MoD culture, continuing optimism bias about program or project budgets, and enduring interservice rivalries despite efforts to promote "multi-domain integration." In addition, there are ongoing struggles to rein in the MoD's cost overruns, enable fungibility and flexibility, and overcome barriers to rapid acquisition and innovation.

Structure of This Report

In Chapter 2, we provide a detailed case study on Australia's defense resource planning. In Chapter 3, we provide a detailed case study on Canada's defense resource planning, and in Chapter 4, we provide a case study on the UK's defense resource planning. In Chapter 5, we review key insights across the three case studies.

---

[44] MoD, *Annual Report and Accounts: 2020–21*, January 20, 2022a, p. 22.

CHAPTER 2

Australia

Andrew Dowse, Benjamin J. Sacks, Austin Wyatt, and Jade Yeung

Australia has a mixed system of government that includes a representative democracy, a constitutional monarchy, and a federation of states. The Australian Constitution defines the government's three branches (executive, legislature, and judiciary) and describes how they share power: The parliament (legislature) makes and changes the law; the executive branch—the reigning government, represented by the prime minister and his or her ministers, who also are elected parliamentarians—puts the law into action; and the judiciary settles disputes about the law.

Within this system, the Australian Department of Defence (hereafter referred to as *Defence*) is responsible for defending Australia and its overseas territories and for executing defense-related missions wherever and whenever required by the national government. The Australian government aligns resources to fill Defence mission needs through its IIP, which is the plan for future capability investment, and its Portfolio Budget Statement, which is the proposed allocation of resources to outcomes.[1] Defence comprises military forces, collectively known as the Australian Defence Force (ADF), as well as policy and support elements. The Defence organization is structured under a diarchy, with a civilian Secretary of Defence and a military officer as Chief of the Defence Force (CDF) who report to the Minister for Defence, an elected parliamentarian from the reigning majority party. In recent decades, Defence has participated in operations in Iraq and Afghanistan, various peacekeeping missions (e.g., in Timor-Leste), and humanitarian assistance and disaster response missions in the Indo-Pacific theater.

The Australian government's *2020 Defence Strategic Update* defines specific defense objectives, including the following:[2]

- prioritizing "our immediate region . . . for the ADF's geographical focus"
- growing "the ADF's self-reliance for delivering deterrent effects"

---

[1] The role of Portfolio Budget Statements is to inform senators and MPs of the proposed allocation of resources to outcomes by each department. These documents also explain and justify those allocations with regard to outcomes in the context of the whole-of-government budget.

[2] Australian Department of Defence, *2020 Defence Strategic Update*, 2020a, p. 25, sec. 2.13.

- expanding "Defence's capability to respond to grey-zone activities, working closely with other arms of Government"
- enhancing "the lethality of the ADF for the sorts of high-intensity operations that are the most likely and highest priority in relation to Australia's security"
- maintaining "the ADF's ability to deploy forces globally where the Government chooses to do so, including in the context of US-led coalitions"
- enhancing "Defence's capacity to support civil authorities in response to natural disasters and crises"
- acknowledging the inability to rely on the previously accepted ten-year strategic warning time frame for major conflict
- emphasizing deterrence as a key outcome, with an expectation that this will require increased offensive capability.

Defence operates in close concert with several allies, especially the United States, and leverages those alliances and partnerships as a central tool of national security. Australia is a member of the Five Eyes security agreement and the Quadrilateral Security Dialogue (the Quad). Australia joined the NTIB in 2017, although the potential of allied cooperation under the NTIB has not yet been realized because of export control barriers.[3] It is also a member of the recently signed AUKUS agreement, which is designed to strengthen the defense relationship among the three countries through technological cooperation and the co-development of advanced capabilities. The strong allied focus of Australia's defense strategy emphasizes the importance of interoperability and, in some cases, integration—a key consideration in acquisition and force generation. As we discuss later, this emphasis also poses a challenge to Australia's ability to independently pursue flexibility. Finally, Australia is a strategically located partner in the Indo-Pacific theater and shares U.S. concerns about China's military rise.

Australia has a highly organized defense budgeting system; its budgeting process is based on a systemic strategy-to-task approach in which there is clear alignment between resources and the outcomes that they deliver. Notably, Australia's budget allocations are expected to rise as it prepares to develop its military to respond to new and emerging threats in the Indo-Pacific.

A key strength of the Australian system is that Defence can lay out its baseline budget over a ten-year period through strategic documents, most recently the *2016 Defence White Paper*.[4] Defence, therefore, enjoys an important degree of budget surety. The unapproved, fungible portions of the IIP (which is reviewed biannually) provide fiscal flexibility among

---

[3] Thomas-Noone, 2019.

[4] Australian Defence official, interview with the authors, October 2022. See Australian Department of Defence, 2016b.

Australia's military services and programs.[5] Allocations for unapproved projects in the IIP can be shifted in biannual reviews to facilitate the introduction, cancellation, prioritization, or deprioritization of capability programs in response to external factors. The smaller and more integrated nature of the ADF allows for investment funding to be shifted across the services—a capability that the United States currently lacks. The IIP provides public transparency and prepares industry through demand signals of potential future requirements.[6]

A key challenge is that the Australian ODCS, like DoD's PPBE System, can take many years (up to a decade) to allow a capability program to pass through the ODCS and the IIP, from unfunded concept to funded reality. Finally, given the close defense cooperation relationship between Australia and the United States, any changes to DoD's PPBE System likely will lead to secondary effects for the Australian system.

Overview of Australia's Defense Budgeting Process

Defence operated with a nominal budget of AUD 48.7 billion (U.S. $34.52 billion) in FY 2022–2023.[7] This budget included funding for the Australian Signals Directorate, a distinct organization within Defence, as well as smaller defense-related agencies and programs.[8] The Australian defense budget equates to more than 2 percent of Australia's GDP.[9] This percentage reflects a continuing commitment to increasing defense spending in response to a worsening geostrategic threat environment. Table 2.1 shows that Australia's largest defense expenditures are for acquisition, sustainment, and workforce.

Australian defense spending is guided by periodic strategic planning documents, such as the *2016 Defence White Paper* and *2020 Defence Strategic Update*,[10] which fulfill a similar function to that of the U.S. NDS and generally follow a regular four- to five-year cycle. These documents are shown in Figure 2.1.

The *2020 Defence Strategic Update* offered a new policy framework for an evolving strategic environment. That update provided the basis for greater urgency in defense planning, which, for the first time, departed from the strategic assumption of ten years' warning before

---

[5] Australian Defence official, interview with the authors, October 2022. The three Australian military services are the Royal Australian Navy, the Australian Army, and the Royal Australian Air Force.

[6] Watt and Brangwin, undated.

[7] Australian Department of Defence, undated-a.

[8] For a more in-depth look at the Australian defense organizational structure, see Australian Department of Defence, 2022a, p. 17; and Australian Department of Defence, 2022c, p. 10, Figure 2.

[9] Marcus Hellyer and Ben Stevens, *The Cost of Defence: ASPI Defence Budget Brief 2022–2023*, Australian Strategic Policy Institute, June 2022.

[10] Nicole Brangwin and David Watt, *The State of Australia's Defence: A Quick Guide*, Australian Parliament, July 27, 2022. See Australian Department of Defence, 2016b; Australian Department of Defence, 2020a.

TABLE 2.1

Planned Defense Expenditures, by Key Cost Category

Serial Number	Cost Category	2021–2022 Actual Result ($m)	2022–2023 Previous Estimate ($m)	2022–2023 Budget Estimate ($m)	2023–2024 Forward Estimate ($m)	2024–2025 Forward Estimate ($m)	2025–2026 Forward Estimate ($m)	Total ($m)
1	Workforce	13,522.0	14,160.0	14,167.1	14,535.3	15,202.1	15,773.3	73,199.8
2	Operations	469.4	193.2	216.5	3.3	1.3	1.3	691.8
3	Capability acquisition program	14,389.8	16,263.5	16,215.3	18,438.8	18,986.6	20,240.1	88,270.6
4	Capability sustainment program	14,386.8	14,975.6	15,065.4	15,353.7	16,380.2	17,214.8	78,400.9
5	Operating	3,260.8	2,387.0	2,388.9	2,449.0	2,259.0	2,240.7	12,598.4
6	Total Defence planned expenditure	46,028.8	47,979.3	48,053.2	50,780.1	52,829.2	55,470.2	253,161.5

SOURCE: Adapted from Australian Department of Defence, *Portfolio Budget Statements 2022–23: Defence Portfolio— Budget Initiatives and Explanations of Appropriations Specified by Outcomes and Programs by Entity: Australian Signals Directorate*, October 2022b, p. 15, Table 4b.

NOTE: These categories are funded by appropriations and own-source revenue. Costs are shown in AUD.

FIGURE 2.1

Regularly Updated Strategic Plans Guide Australia's Defense Resource Decisions

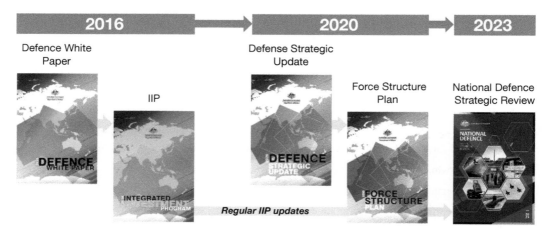

SOURCE: Australian Department of Defence, "Strategic Planning," webpage, undated-c.

a major conflict. After assuming power in 2022, the new Labour government (led by Prime Minister Anthony Albanese) endorsed the principles of the update, reflecting the general bipartisanship around defense matters in Australian politics. But Labour also recognized major challenges for the defense budget, including affordability (which is associated with a weaker Australian dollar relative to the U.S. dollar); the need for a stronger ADF, given the deteriorating strategic environment (and an accompanying increase in needed resources); and broader budget pressures. The interim budget delivered by the new government in October 2022 maintained military spending just above 2 percent of GDP but did not specify any increases. In August 2022, the Australian government initiated a substantial review to assess whether Australia had the necessary defense capability, posture, and preparedness, given the strategic circumstances. The unclassified report of the Defence Strategic Review (DSR) was publicly released on April 24, 2023, and addressed aspects of Australian PPBE-like processes.[11]

Before delving into the specifics of Defence's budgeting process, it is worth reviewing the particulars of the Australian legislative and executive branches. Australia is a parliamentary constitutional monarchy modeled on the UK's (the Westminster system) with a bicameral parliament comprising the House of Representatives (lower house) and Senate (upper house). The government is formed by the party with the majority of seats in the lower house, although minority governments can be formed with the support of minor parties and independent members if neither major party secures a majority in a general election. As a result, the government will structurally hold a majority in the lower house, and, when the party in power does not hold a Senate majority, it will negotiate with the upper house to secure the passage of legislation, including on matters of confidence and supply (i.e., the budget).

Australian federal government elections are held at least every three years. The two major parties in Australia take a relatively bipartisan approach to defense; hence, a change of government does not necessarily result in any significant change in defense plans or budget allocations. New governments sometimes direct the department to begin work on a new defense white paper; however, such changes in strategic guidance are typically related more to changes in the geostrategic environment than to politics.[12]

It is also important to acknowledge that the Australian electorate votes for parties, not individual prime ministers. Prime ministers are selected by the party that holds the majority in the new government, and, subsequently, the prime minister appoints senior elected colleagues to ministerial positions—comparable to the appointment of secretaries in the U.S. cabinet. Each minister is therefore an elected MP or senator and assumes responsibility and accountability for a given department's functions. Under the Minister for Defence, there is both a departmental secretary, the professional head of Defence (a career bureaucrat rather

---

[11] The DSR is more likely to reflect changes in investment and program priorities than an overall change in the budget allocation.

[12] The 2022 change in government leadership from the Liberal to the Labour party did not result in any substantial change in priorities or budget for Defence, and the new government did not initiate a new defense white paper.

than a political appointee), and the CDF, a military officer.[13] Overall, the legislative and executive arms of the Australian system are more closely linked than they are in the United States.

Generally speaking, the Australian Parliament has used the same legislative process since the federation and the country's independence from the UK in 1901. Each year, the House of Representatives debates and approves appropriations bills (which contain executive budgets, including for defense) and submits them to the Senate for review and approval. After Senate approval, the bills are sent to the governor-general (whose duties include serving as commander in chief of the ADF) to secure royal assent.

Budget planning commences in September and October of each year.[14] Budget development and the costing of new policy proposals occur between December and February.[15] At this point, departments hand over primary responsibility to the Treasurer of Australia, who oversees the development of a draft whole-of-government budget. MPs and senators are typically able to review this draft beginning in February. The prime minister and cabinet (i.e., the executive) submit the draft budget, along with each department's Portfolio Budget Statement, to parliament for review in March.[16] The budget is formally introduced in the House of Representatives on budget night, which is normally the second Tuesday in May.[17] After the opposition leader's right of reply, Appropriation Bills Nos. 1 and 2 undergo several weeks of debate.[18] In the interim, the brief *Particulars of Certain Proposed Expenditure* document moves forward to the Senate so that members can get a head start on their review.[19] See Figure 2.2 for an overview of Australia's budget process.

The relevant Senate committee—in this case, Foreign Affairs, Defence and Trade—reviews the proposed defense budget document throughout May before voting.[20] Prior to 1969, the Senate did not use committees to examine specific portions of the budget. Rather, senators collectively examined the entire budget, "address[ing] extremely detailed questions about proposed expenditure to the minister in charge of a portfolio or to a minister repre-

---

[13] The CDF is Australia's senior military officer, the only four-star officer in the ADF. The CDF leads the integrated Australian Department of Defence and ADF as a diarchy with the Defence Secretary.

[14] Australian Parliamentary Budget Office, "Overview of the Budget Process," fact sheet, 2022.

[15] Australian Parliamentary Budget Office, 2022.

[16] Australian Government, "Portfolio Budget Statements," webpage, undated; Australian Department of Defence, 2022b, p. ix. As noted earlier, the role of Portfolio Budget Statements is to inform senators and MPs of the proposed allocation of resources to outcomes by each department. These documents also explain and justify those allocations with respect to outcomes in the context of the whole-of-government budget.

[17] Daniel Weight and Phillip Hawkins, *The Commonwealth Budget: A Quick Guide*, Australian Parliament, May 7, 2018.

[18] Weight and Hawkins, 2018; Australian Department of Defence, 2022b.

[19] Weight and Hawkins, 2018.

[20] Australian Senate, "Consideration of Estimates by the Senate's Legislation Committees," Senate Brief No. 5, January 2023.

FIGURE 2.2

Australia's Budget Process

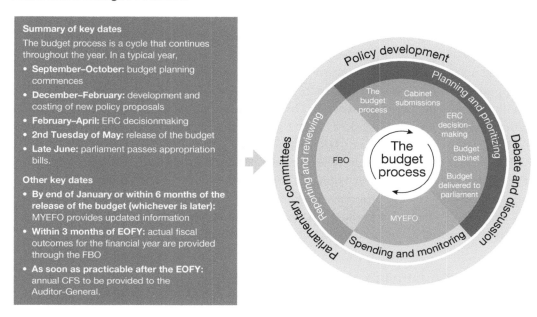

Summary of key dates

The budget process is a cycle that continues throughout the year. In a typical year,

- **September–October:** budget planning commences
- **December–February:** development and costing of new policy proposals
- **February–April:** ERC decisionmaking
- **2nd Tuesday of May:** release of the budget
- **Late June:** parliament passes appropriation bills.

Other key dates

- **By end of January or within 6 months of the release of the budget (whichever is later):** MYEFO provides updated information
- **Within 3 months of EOFY:** actual fiscal outcomes for the financial year are provided through the FBO
- **As soon as practicable after the EOFY:** annual CFS to be provided to the Auditor-General.

SOURCE: Adapted from Australian Parliamentary Budget Office, 2022.
NOTE: CFS = Consolidated Financial Statements; EOFY = end of fiscal year; ERC = Expenditure Review Committee; FBO = Final Budget Outcome; MYEFO = Mid-Year Economic and Fiscal Outlook.

senting a minister in the House of Representatives."[21] By the Australian Parliament's own admission, this was considered to be a slow, laborious, and cumbersome process.[22] In 1969, the Senate delegated budget examination to eight committees, including the Foreign Affairs, Defence and Trade Committee. The budget is formally passed in late June,[23] and the same calendar is used for all departments.

The minority party can participate in this public debate and can leverage the biannual Senate Estimates process to question departmental leadership; however, the budget has historically progressed unimpeded through the lower house. Furthermore, although MPs and senators are not legally barred from requesting information from executive departments, there are procedural and normative barriers. Australian Public Service employees (federal civil servants) are instructed to refer all inquiries from legislators to designated ministerial liaison sections within their departments. In addition, each minister's office maintains a staff of departmental liaison officers.

---

[21] Australian Senate, 2023, p. 2.

[22] Australian Senate, 2023.

[23] Australian Parliamentary Budget Office, 2022.

Deadlock between the houses is mitigated by the prospect of double dissolution; *double* refers to both houses of parliament. This process is triggered if a bill (including the budget) is passed by the House of Representatives but rejected twice by the Senate. The governor-general, representing the monarchy, has the power to dissolve both houses of parliament and call a general election at the earliest opportunity. The new government must then pass the contested bill in the House of Representatives, and the bill must then be passed to the Senate for approval. If the process is still unsuccessful, the governor-general convenes a joint sitting of both houses to pass the contested bill. This dissolution has occurred seven times since federation and Australia's independence from the UK in 1901,[24] although in none of these cases was the defense budget the cause of that dissolution.

The passage of appropriations bills each year provides authorization for the expenditure of funds for that year. However, unlike in the United States, the annual budget associated with existing services and programs appears in a separate appropriations bill from that for new programs,[25] making it unlikely that existing government services will be blocked and effectively eliminating any need for a continuing resolution.

Specifics of Australia's Budget Mechanisms

Defence recently completed the transition to accrual budgeting,[26] whereby the department submits a budget request for funding to cover ongoing costs. This funding cannot be carried over to the next fiscal year, and the acquisition of in-year funding is based on aggregate spending in the current year. Since 2020, Defence has expressed this spending on a net cash funding basis in its budgeting and reporting. This transition has increased the transparency and accountability of Defence funding and has made it easier to compare financial performance data between departments. The transition also has an important role in Australia's ability to modernize its defense funding process, including producing Portfolio Budget Statements.

Defence has five key cost categories, which are similar to U.S. appropriation categories: workforce, operations, capability acquisition program (including R&D), capability sustainment, and operating costs.[27] There is limited movement among categories, but there is flexibility for "unders" and "overs," meaning that funds can be shifted from categories with a surplus to categories with a deficit. In effect, Defence is given capital appropriation injections to fund major capability acquisitions that are generally outlined in the Portfolio Budget State-

---

[24] Australian Parliament, "Double Dissolutions," webpage, undated.

[25] For details on the separation of appropriations bills for continuing services and new policies, see Adam Webster, "Explainer: Can the Senate Block the Budget?" webpage, The Conversation, May 19, 2014.

[26] Australian Defence official, interview with the authors, November 2022.

[27] In this context, *operating* relates to the forecasted costs to support defense systems, including training on those systems, whereas *operations* relates to nonforecasted costs associated with deployed forces.

ment delivered to parliament. Projects are funded and managed on a whole-of-life basis,[28] accounting for both capital and operating costs.

Defence operates under a "no-win, no-loss" mechanism for operational commitments (i.e., deployments).[29] In other words, Defence is reimbursed for most operational costs and must return unused funds to the treasury. Defence absorbs some level of its costs, but the majority is offset by government reimbursement.

Rapid inflation is an emerging concern that poses a significant risk to Australia's government—a risk that is expected to be addressed in the 2023 Defence Strategic Review. Sharp increases in costs across all colors of money, in addition to strategic drivers for increased defense spending,[30] are putting pressure on the defense budget. While the impact of inflation on GDP is uncertain, the Australian government will have difficult budgeting decisions ahead, despite an indication that Australian defense spending will rise to 2.2 percent of its GDP.[31]

Australia's defense funding for the current fiscal year is allocated through the annual budgetary process and based on the defense white papers and strategic updates that the government releases to the public every four years or so, along with classified planning guidance. The budget provides top-line defense funding certainty by setting forward estimates with a high level of confidence (but not without some uncertainty) for the next three fiscal years. Moreover, defense strategic guidance documents and the IIP provide provisional (or medium-level-of-confidence) funding for ten years. This three-tiered funding stream—current year, top line, and indicative—evolved from a planning approach outlined in the 1976 *Australian Defence* white paper, which instituted a five-year funding allocation for the acquisition program.[32] This approach had been criticized for its lack of strategic direction on long-term procurement decisions.[33] The planning and budgeting approach evolved through subsequent defense white papers, leading to the current process in which the defense budget is baselined over a period of up to ten years, with strong confidence in funding in forward estimates and relative confidence in funding availability over the remainder of the decade. Changes within that period are typically limited to indexation and government decisions about what to add,

---

[28] Under the ODCS, approval to acquire new weapon systems requires an estimate of total costs through the system's projected end of life, meaning that personnel, operating, and sustainment costs are identified in ongoing budgets.

[29] Not to be confused with the day-to-day running of the ADF.

[30] This increased spending includes a significant commitment to expenditure on nuclear-powered submarines; see Lewis Jackson, "Australia's Nuclear Submarine Plan to Cost up to $245 billion by 2055—Defence Official," Reuters, March 14, 2023.

[31] See Minister for Defence Marles' comments in Richard Marles, "Television Interview, ABC News Broadcast," transcript, March 15, 2023.

[32] Australian Department of Defence, *Australian Defence*, November 1976.

[33] Watt and Brangwin, undated.

cancel, prioritize, or deprioritize in the IIP. Defence therefore enjoys some level of budgetary certainty going into each new financial year.[34]

Defence does not use one distinct budgeting process for certain functions—for example, for RDT&E—and a different process for other functions, such as sustainment, operations, and procurement. However, the funding does come from different pools. Funding for defense acquisition, for example, is articulated in the IIP, which provides a ten-year plan for capability investment derived from strategic objectives.[35] Defence prepares its strategic guidance and the IIP, which are reviewed and approved by the executive branch, primarily the Minister for Defence. Accordingly, the IIP is reviewed and adjusted biannually.[36] The Vice Chief of the Defence Force (VCDF) manages the IIP, gathering input from stakeholders across the service branches and joint strategic planning units, such as the Force Design Division. Major acquisitions come through the IIP; the acquisitions are funded through the agencies that draw funds from the approved budgets for each acquisition. Such draws and expenditures of acquisition funds occur at a rate that is aligned with the approved overall Defence expenditure for a given year.

The IIP includes both approved government projects and unapproved, fungible programs that can be shifted "left" or "right" (i.e., accelerated or delayed) as needs arise.[37] The IIP informs the budget document that is submitted to parliament for approval and is highlighted in the Portfolio Budget Statement. To manage the risk of underachievement (or overexpenditure) relative to the acquisition budget, the IIP is 20-percent overprogrammed for acquisition in the current financial year.

Funding for operations, sustainment, and personnel is separate from the IIP. These funds are allocated through the regular government budgetary process, the defense portion of which is explained in Defence's Portfolio Budget Statement.

Defence has been considering changes to the IIP to increase agility in force structure in view of the pacing threat; potential adversaries may update their capabilities on shorter timelines. This concern was a key tenet of the *2020 Defence Strategic Update*.[38] The deterioration in Australia's strategic circumstances has led to calls for a more expeditious process that prioritizes early capability over current, slower processes that are focused on risks, openness, fairness, and value for money. Such updated views may not align with the Australian govern-

---

[34] Australian Defence official, interview with the authors, October 2022.

[35] The IIP includes required acquisition plans for the following ten years with year-by-year funding breakdowns (acquisitions typically overprogrammed in the current fiscal year). But it also forecasts major acquisitions out to 20 years.

[36] Australian Defence official, interview with the authors, October 2022.

[37] Australian Defence official, interview with the authors, October 2022; Australian Department of Defence, *Department of Defence Annual Report 2019–20*, 2020c.

[38] Australian Department of Defence, 2020a.

ment's acquisition performance concerns,[39] although such concerns might be overstated.[40] The trade-off between speed and performance may be the exception—for urgent operational requirements—rather than a general rule.

Inadequate agility in developing the future force is a key challenge for Defence's PPBE process. There is a cultural aversion to acquisition risk within Defence, and that lengthens review times and holds up funds that could be spent on other projects.

The One Defence Capability System and Its Defence Capability Assessment Program

The current system and culture are reflected in Australia's acquisition system: the ODCS. The ODCS was the end point of a series of government responses to the 1998 Joint Standing Committee Report *Funding Australia's Defence*, which noted that Defence was unique in being largely exempt from efficiency demands and operating on a "global Defence budget" that gave the Minister for Defence and service chiefs significant discretion on how top-line funds were allocated.[41] Established in 2015 as the most recent in a string of measures to increase government budgetary oversight, the ODCS was intended to further centralize Defence's PPBE process and enhance contestability reviews. The current system is therefore stricter in accounting for how money is spent throughout the budgetary process. As a result, it can take six to ten years for an acquisition project to be introduced into service.

Stage 1 of the ODCS Process

The ODCS is a four-stage process that begins with an assessment of strategy and concepts (see Figure 2.3). In the first part of stage 1, Defence strategy documents aim to articulate how defense strategies can be realized and effectuated as operational concepts. The concepts can be incremental, novel, or experimental, but they must align with either (1) one or more of 35 capability programs (including ten joint programs that span Australia's five warfighting domains)[42] or (2) one or more of the 11 multidomain programs, which involve a greater degree of coordinated development across capability programs and domains.

In the second part of the strategy and concepts stage, proposed concepts enter the Integrated Force Design Process. The primary feature of this process is the Defence Capability Assessment Program (DCAP). The DCAP, which theoretically operates on a continual two-year cycle, "is the main analytical process for converting strategic priorities into an expression of the intended set of capabilities that provide [an] optimised range of options for addressing

---

[39] Daniel Hurst, "Defence Projects Suffer $6.5bn Cost Blowout as Marles Promises More Scrutiny in Future," *The Guardian*, October 9, 2022.

[40] Marcus Hellyer, "The Real Costs of Australia's Defence Budget 'Blowout,'" webpage, The Strategist, October 18, 2022.

[41] Australian Parliament, Joint Standing Committee on Foreign Affairs, Defence and Trade, *Funding Australia's Defence*, May 8, 1998, p. 27.

[42] These domains are maritime, land, air, space, and information and cyber.

FIGURE 2.3

One Defence Capability System

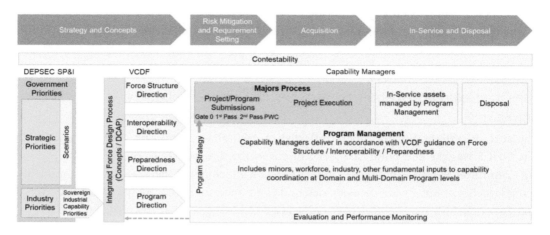

SOURCE: Reproduced from Australian Department of Defence, *Defence Capability Manual*, version 1.1, December 3, 2021a, p. 5, Figure 1-3. Used with permission.

NOTE: DEPSEC SP&I = Deputy Secretary Strategy, Policy, and Industry; PWC = Parliamentary Standing Committee on Public Works.

strategic risks."[43] The DCAP is also the key activity that promotes agility, but it is currently linked to government updates of strategic guidance, which may not be sufficiently agile.[44] There has been an effort to make these updates more frequent and ongoing, as articulated in the 2020 *Defence Strategic Update*.[45]

Our interviewees noted the importance of the DCAP process in not only adding or revising investments to address emerging threats but also eliminating extant programs that will no longer be relevant in the future threat environment. Eliminating extant programs is an ADF challenge that the current Defence Strategic Review is addressing, among other things. Because Australia has a smaller force than the U.S. military, for example, it is comparatively easier for the ADF to change its structure. Its smaller size could offer an advantage in terms of flexibility or a disadvantage if the capabilities and skills to address enduring threats need to be relinquished to focus on emerging threats.

The DCAP, which employs a capability-based planning approach "to make planning more responsive to uncertainty, economic constraints, and risk,"[46] nonetheless has its own circu-

---

[43] Australian Department of Defence, 2021a, p. 29.

[44] Australian Defence official, interview with the authors, October 2022.

[45] Australian Department of Defence, 2020a; Australian Defence official, interview with the authors, October 2022.

[46] Australian Department of Defence, *Force Design Guidance*, version 1.1, December 3, 2021b, p. 4. A capability-based planning approach is distinguished from scenario-based planning by its focus on internal resources as a mechanism for achieving desired outcomes.

lar eight-step process (see Figure 2.4). The DCAP first identifies "which force packages are affected by change, to what extent, and at what time."[47] Next, it quantifies the proposed capability program's risk by "develop[ing] risk statements for each joint capability effect using the ADF risk framework."[48] Third, it prioritizes risk mitigation by running workshops and discussing potential risks with Defence's internal Investment Committee, which is led by the VCDF. Fourth, it develops capability program options. Fifth, it tests those options, "looking out for twenty years."[49] Sixth, the DCAP identifies offset strategies for new investments. Seventh, it tests the capability portfolio options "to confirm net positive impact."[50] Finally, a decision is made as to whether to incorporate an identified, quantified, prioritized, developed, tested, offset, and retested capability program into the IIP as a nominee for gate 0 review by

FIGURE 2.4
Defence Capability Assessment Program Process

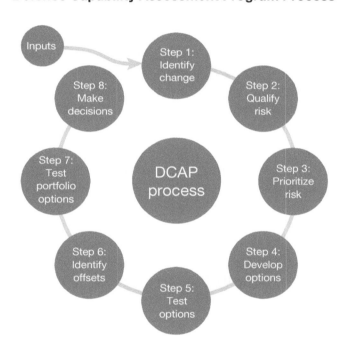

SOURCE: Adapted from J. Boyce, N. Tay, and C. Row, "Consideration of Enabling and Enterprising Functions Within Defence Force Design," *Proceedings of the 24th International Congress on Modelling and Simulation*, December 2021, p. 919.

---

[47] Australian Department of Defence, 2021b, p. 8.

[48] Australian Department of Defence, 2021b, p. 8.

[49] Australian Department of Defence, 2021b, p. 8.

[50] Australian Department of Defence, 2021b, p. 8.

the Investment Committee.[51] This gate 0 entry into the IIP will establish expectations for the program in terms of its scope of the capability, expected schedule, and cost.

The eight-step DCAP process results in proposed capability programs for further review in the ODCS. The candidate capability programs are organized into sets of fundamental inputs to capability (FICs). These FICs are similar in scope to the U.S. doctrine, organization, training, materiel, leadership and education, personnel, and facilities (DOTMLPF) set of characteristics that collectively constitute a capability. Australian FICs include organization, command and management, personnel, collective training, major systems, facilities and training areas, supplies, support, and industry.[52]

Throughout the DCAP process, which is still part of stage 1 of the ODCS process (under the VCDF column in Figure 2.3), the ODCS also engages with the contestability framework,[53] which is a group of evaluators who are responsible for "providing independent review of capability proposals to ensure they are aligned with strategy and resources and can be delivered in accordance with Government direction."[54] Once a capability program has exited the DCAP process, the program is assigned to the relevant capability manager. The capability managers are the chiefs of the three services, the Chief of Joint Capabilities, the Chief Defence Scientist, the Chief of Defence Intelligence, the Chief Information Officer, and the Deputy Secretary Security and Estate.[55] The capability programs (packaged into their respective FICs) are then represented in the IIP as the DCAP's primary output.[56] The DCAP process also leads to defense workforce proposals, an estate plan, and capability program strategies.[57] The IIP ultimately informs the budget submitted to parliament.

Stage 2 of the ODCS Process

The second stage of the ODCS—risk mitigation and requirement-setting—involves internal gate approvals and government approvals. On average, it takes two years for each capability program to progress between gates 0 and 1 and an additional two years to progress between gates 1 and 2, both of which are in stage 2 of the ODCS process. The IIP is a central document in this stage.

Capability proposals for gate 1 require an interim business case for consideration by the Investment Committee, followed by the final submission of a capabilities packet for government approval. The content of this business case would depend on the nature of the capabil-

---

[51] Australian Department of Defence, 2021b, p. 8.

[52] Australian Department of Defence, 2021a, pp. 12–13.

[53] For background on contestability in Defence, see Cynthia R. Cook, Emma Westerman, Megan McKernan, Badreddine Ahtchi, Gordon T. Lee, Jenny Oberholtzer, Douglas Shontz, and Jerry M. Sollinger, *Contestability Frameworks: An International Horizon Scan*, RAND Corporation, RR-1372-AUS, 2016.

[54] Australian Department of Defence, 2021a, p. 24.

[55] Australian Department of Defence, 2021a, pp. A-2–A-3.

[56] Australian Department of Defence, 2021b, p. 5.

[57] Australian Department of Defence, 2021b, p. 5.

ity, its cost, and required timeline; for example, a novel major capability acquisition would require a far more extensive business case. The final capabilities packet includes the formal submission to an appropriate cabinet committee, the business case, a joint capabilities needs statement, a capability workforce plan, a test and evaluation master plan, a capability definition document, and other supporting documents.[58] The government delegate reviewing the packet determines whether the capability program "relates to strategic priorities and the range of capability options."[59]

If the government delegate—ranging from a single minister to the National Security Committee of Cabinet (NSC)—approves this first pass, the packet undergoes further refinement to clarify documentation and costs (which might require engaging with industry) between gates 1 and 2 for the additional two years, on average. Industry is involved to a greater extent ahead of the second pass to provide high-confidence assessments of costs, schedules, and risks. Capability managers also undertake capability realization planning at this point.

Next, the capability proposal proceeds with its detailed business case to gate 2, where it is reviewed again by the Defence Investment Committee before a second-pass consideration by the government. If the government approves the capability proposal on the second pass, the proposal shifts from unapproved to approved investment funding, allowing acquisition agencies to draw down funding. Capability managers (i.e., service chiefs) are allocated their resources for approved projects and are accountable for the required capability outcomes. The managers have the flexibility to move resources within their organizations to meet those capability outcomes, with limitations in relation to moving funds between buckets of money.

To reiterate, Australian processes for risk mitigation and requirement-setting typically presume at least two years to achieve approval of each pass; hence, new capabilities often take up to a decade (including acquisition timelines) to be introduced into service. However, certain high-priority or low-risk capabilities can undergo an accelerated combined first-second (or simply combined) pass, which somewhat reduces the overall timeline.

At each stage, the capability proposal is examined, through the contestability function, for the answers to the following:

- Do the requirements and resultant capabilities align with the articulated ADF strategy and agreed-on resources?
- Does the capability provide value for money?
- Is it the best option among a set of potential alternatives?
- What is its implementation plan, and how will risk be managed?[60]

---

[58] Australian Department of Defence, 2021a, p. 42.

[59] Australian Department of Defence, 2021a, p. 40.

[60] Australian Department of Defence, 2021a, p. 41.

If a capability package makes it through the second pass and the contestability review, Defence will engage with industry and coordinate with other FIC providers ahead of acquisition.[61]

Stage 3 of the ODCS Process

In the third stage, acquisition, the IIP's capabilities programs are translated into the budget and are acquired by Defence through the approved defense budget. Although capability managers ultimately remain responsible,[62] they normally delegate their responsibilities to the lead delivery group, which is consistent with the defined capability, schedule, and cost criteria for the acquisition. Capability managers are also responsible for the FICs that relate to the other three buckets of money (sustainment, personnel, and operating costs).[63] Acquisition requires "appropriate management oversight"; "control measures to detect variations from budget, schedule, scope, and other aspects of approved plans"; and risk management.[64]

Stage 4 of the ODCS Process

In the fourth stage, in-service and disposal, capability managers maintain oversight over the capability program during its service life and during its eventual disposal.[65] Sustainment may include adjustments to the program, such as if its FICs change or if its status is altered as a result of political decisions or for administrative reasons.[66] Defence uses this process for all capability program functions.

The ODCS process is relatively new. The current defense budgeting process is influenced by the Financial Management and Accountability Act of 1997; the Charter of Budget Honesty Act of 1998; and the Public Governance, Performance, and Accountability Act of 2013. The First Principles Review of 2015 established the ODCS process, including the contestability framework, to streamline and centralize the defense budget process and to enhance its transparency and value for money.

Decisionmakers and Stakeholders

The IIP—and, ultimately, the budget—must be designed to align with the defense documents that lay out the Australian government's defense priorities, most recently the *2016 Defence*

---

[61] Australian Department of Defence, 2021a, p. 43.

[62] For example, "[a] senior Defence officer (typically 3-star or [Senior Executive Service] Band 3) accountable for management of subordinate Capability Programs and oversight of any assigned Multi-Domain programs, including the development, delivery, introduction, preparedness and withdrawal of capabilities, in accordance with Defence policy and procedures" (Australian Department of Defence, 2021a, p. A-2).

[63] Australian Defence official, interview with the authors, October 2022.

[64] Australian Department of Defence, 2021a, p. 44.

[65] Australian Department of Defence, 2021a, p. 48.

[66] Australian Department of Defence, 2021a, p. 49.

White Paper and *2020 Defence Strategic Update*.[67] Whereas such strategic guidance may be prepared by Defence, it is approved by the Minister for Defence, who is a parliamentarian and a senior cabinet member. At the top of the government, defense matters fall to the NSC, which includes the prime minister, deputy prime minister, Minister for Defence,[68] treasurer, Minister for Finance, and other ministers when their portfolios are involved.[69]

Capability-related submissions are examined first, however, by the Minister for Finance–led National Security Investment Committee of Cabinet. The top-of-government NSC is responsible for approving all National Security Investment Committee decisions and is the ultimate decisionmaker for all capability program decisions.[70] The NSC also considers strategic priorities and operational matters, making it the peak decisionmaking authority for Australian defense budgets.

At the department level, decisionmaking for resources is undertaken through the Defence Committee, which comprises the CDF, Defence Secretary, VCDF, Associate Defence Secretary, and Chief Finance Officer (CFO).[71] Much of the management of budgeting is undertaken between the VCDF and Associate Secretary of Defence, who effectively represent the resource demand and supply sides. Additionally, the VCDF chairs the Investment Committee, which makes departmental decisions associated with the execution of the IIP. The VCDF delegates significant program authority to capability managers, who are responsible for "capability realisation."[72] Capability managers, in turn, delegate to lead delivery groups the responsibility "for coordinating and integrating the FIC[s] on behalf of the Capability Manager."[73] Ultimately, however, decisions concerning the IIP, new capability programs, and the budget before submission to parliament are the responsibility of the civilian executive government—the prime minister and cabinet.

Budgets are allocated and managed at the department level and reflect overall priorities, input from service components and related organizations, and baseline personnel and operating costs. Although the service components and related organizations have some flexibility in the current fiscal year, they are expected to expend resources and deliver outcomes that are consistent with the Portfolio Budget Statement. Thus, there is centralized priority-setting but decentralized execution. Occasionally, there is a need to negotiate transfers of funding within the current fiscal year in response to changing priorities.

---

[67] See Australian Department of Defence, 2016b; and Australian Department of Defence, 2020a.

[68] The current Minister for Defense is also the deputy prime minister.

[69] Australian Department of Defence, 2021a, p. 14.

[70] Australian Department of Defence, 2021a, p. 14.

[71] The three service chiefs are no longer members of the Defence Committee following the 2016 removal of their statutory appointments.

[72] Australian Department of Defence, 2021a, pp. 10, 12.

[73] Australian Department of Defence, 2021a, p. 12.

The government formally interacts with industry in at least three stages during the budgetary process. First, while developing strategies and concepts (in stage 1 of the ODCS), Defence engages with industry in an effort to maximize Australia's control over "the skills, technology, intellectual property, financial resources and infrastructure that underpin the [Sovereign Industrial Capability Priorities]."[74] At this stage, the government also provides industry with visibility into future acquisition opportunities in the IIP through its publication of the *Defence Industrial Capability Plan*.[75]

Second, during the risk mitigation and requirements phase (in stage 2 of the ODCS), as capability programs progress toward gate 2, Defence engages with industry to assess options to acquire new capabilities.[76]

Third, Defence interacts with industry through the Defence Science and Technology Group and the Defence Innovation Hub (DIH). When the technology readiness level (TRL) is low,[77] it is cultivated by the Defence Science and Technology Group through the Next Generation Technologies Fund.[78] More-mature technologies (TRLs 5–8) can attract funding from the DIH, Defence Science and Technology Group capability programs, and the Defence Materials Technology Centre. As the primary steward for such projects, the DIH received AUD 100 million (U.S. $69 million) per year "to support innovative projects from industry, which comes out of the IIP."[79] An interviewee explained that DIH projects "seek to draw down funds for the forward estimates" in October, and this funding "then proceeds through two minister approvals." The funding is "tracked through the normal budget process," with priorities managed at the one-star level.[80] The Defence Investment Committee ultimately approves DIH projects.

There is no formal interaction between the executive branch and the legislative branch before the budget is presented to parliament. However, the ministers remain responsible to the parliament both in their executive branch roles and as parliamentarians, which includes ensuring that the legislative branch can access budgetary information for accountability purposes. For example, during a daily question time session in the House of Representatives, ministers are obligated to answer questions about their budgets from members (including from the opposition party).[81] Similarly, the Senate, in its review role, asks questions of

---

[74] Australian Department of Defence, 2021a, p. 27. For more on these capabilities, see Australian Department of Defence, "Sovereign Industrial Base Capability Priorities and Plans," webpage, undated-b.

[75] Australian Department of Defence, *2018 Defence Industrial Capability Plan*, 2018.

[76] Australian Department of Defence, 2021a, p. 43.

[77] The TRL scale is 1–9.

[78] Australian Defence official, interview with the authors, October 2022.

[79] Australian Defence official, interview with the authors, October 2022.

[80] Australian Defence official, interview with the authors, October 2022.

[81] Australian Parliamentary Education Office, "Question Time in the House of Representatives," fact sheet, 2022.

Defence officials in relation to budgeting priorities, performance, and issues with biannual estimating activities.

Unlike in the United States, Defence does not provide a list of unfunded programs to the legislature. Rather, parliament has visibility of the IIP, and funded programs for a given year are highlighted in the Portfolio Budget Statement. The legislature never forces earmarks to ensure that Defence funds a certain item. The prime minister and the Minister for Defence can influence investment priorities, however, even overruling the department. One example was when the Minister for Defence decided to invest in F/A-18 *Super Hornets* without seeking advice from Defence.

The time to production for IIP programs is, to an extent, determined by industry partners, not Defence. One view that arose during our interviews was that the IIP is "driven by what the suppliers are able to deliver."[82]

Planning and Programming

Currently, budgetary management is the responsibility of capability managers, and budgetary pressures or issues are reported by Defence's 11 groups and services to their respective finance officers. These officers report to their group or service heads, but they are also accountable to the Defence CFO. The Defence Finance and Resources Committee, a subcommittee of the Defence Committee, addresses financial matters. Whereas each group and service has in-year allocations that are consistent with the Portfolio Budget Statement, funding transfers to meet unforeseen or emerging priorities may be undertaken through the Defence Committee or its subcommittee.

Budgetary needs for future capabilities are developed through the ODCS process and involve industry engagement and FIC analysis by capability managers. Should the budget for proposed capabilities at gate 2 exceed the unapproved (gate 0) allocation within the IIP, the Defence Investment Committee may consider reducing the basis for provisioning the capability (typically by reducing the number of systems) or increasing the IIP allocation, which would require reprogramming the IIP and result in other programs being correspondingly reduced or delayed.

When forming a defense budget, the cabinet and Defence rely on multiple strategic documents that outline planning and program requirements. These documents include the *2016 Defence Industry Policy Statement*, *2016 Defence White Paper*, the *2017 Strategy Framework*, 2019 *Defence Policy for Industry Participation*, *2020 Defence Strategic Update*, and *2020 Force Structure Plan*.[83] They also refer to the classified Defence Planning Guidance and CDF Pre-

---

[82] Australian Defence official, interview with the authors, October 2022.

[83] Australian Department of Defence, *2016 Defence Industry Policy Statement*, 2016a; Australian Department of Defence, 2016b; Australian Department of Defence, "Part 3: Government Direction," *The Strategy Framework*, 2017, pp. 5–7; Australian Department of Defence, *Defence Policy for Industry Participation*, 2019; Australian Department of Defence, 2020a; Australian Department of Defence, *2020 Force Structure Plan*, 2020b.

paredness Directive; these higher-level documents inform the department's strategic and budgetary planning, particularly by helping the cabinet and Defence determine which capabilities to add to the IIP and, thus, determine the IIP's contribution to the government's annual budget. The publicly available strategic guidance documents are updated as directed by the Australian government, typically every four years. Internal documents, such as the classified Defence Planning Guidance and CDF Preparedness Directive, are updated annually.

The Portfolio Budget Statement is the principal document for explaining the links between plans and budgets. An explanatory document for policymakers and the public, the Portfolio Budget Statement outlines Defence's strategic and resource direction, budget expenses and outcomes, and funding levels for the next year for the top 30 military acquisition programs. Because it focuses on the current fiscal year, the Portfolio Budget Statement includes only IIP projects that receive funding in the current fiscal year.[84]

For the IIP, major decisions are made at several points in the ODCS process. This includes Defence decisions at gates 0, 1, and 2 and first- and second-pass approvals by the government. During these checks, Defence and the government address whether the proposed capability programs meet contestability, risk, and Sovereign Industrial Capability Priorities criteria; the latter is intended to "maximise Australian industry involvement."[85] In addition, the programs are examined to see whether they offer value for money, manageable risk, and the best of a selection of options.[86] Depending on the size of the investment, second-pass approval may be given by two ministers (usually Defence and Finance) or by the NSC. In approving the overall defense budget, parliament is the ultimate authority.

Defence uses cost modeling to develop the defense budget estimate: "This included assessing the ADF's ability and capacity in relation to a range of possible scenarios, including responding to natural disasters, and managing a number of concurrent tasks."[87]

Decisionmakers at gates 0, 1, and 2 are guided by the Force Design Division, a unit under the VCDF's purview; the Joint Capability Needs Assessment, which counteracts service stovepiping; a project execution strategy; and a gate 0 business case. Another tool that Defence uses to reach decisions is committee reviews, both within the department (e.g., the Investment Committee) and at the whole-of-government level (e.g., the Expenditure Review Committee). Defence recently introduced in-year forecasting to better control the budget and prevent overspending within individual programs; after three years of this forecasting, Defence still

---

[84] See Australian Department of Defence, 2022b, p. xi.

[85] Australian Department of Defence, undated-b.

[86] Australian Department of Defence, 2021a, p. 41.

[87] Australian Department of Defence, 2020a, p. 34; Australian Defence official, interview with the authors, October 2022.

faced a 1.2-percent overrun (AUD 535 million, or U.S. \$336.98 million) in FY 2020–2021.[88] As a result, it continues to refine this process.[89]

Some projects may be supported by the Defence Science and Technology Group's Next Generation Technologies Fund and the DIH as part of their development processes. These funding initiatives help raise the TRL of potential systems but do not guarantee that they will be subsequently funded. In fact, the operationalization and commercialization of new systems by Australian industry has suffered in the past in a "valley of death" between development and acquisition. Defence has attempted to mitigate this issue through the Next Generation Technologies Fund in certain areas and by funding only DIH proposals that align with programs in the IIP. However, this approach could lead Defence to fund only known applications for technologies, potentially hindering innovation—and capability development and acquisition processes could still delay the adoption of these technologies. Two interviewees recognized these ongoing problems but noted that there is an increasing appetite to make exceptions to fast-track truly game-changing technologies.[90]

Australia's R&D investment, while regionally competitive, remains significantly lower than that of the United States.[91] This difference is highlighted by the fact that DoD spent \$92 billion on R&D in 2022,[92] while the entirety of Australia's defense budget that year was approximately \$32 billion.[93]

On a national level, total Australian investment in R&D equated to 1.79 percent of GDP in 2020;[94] the government contribution was roughly 0.56 percent of GDP.[95] By comparison, the U.S. federal government investment in R&D equated to 3.39 percent of GDP in 2020.[96] In dollar terms, this represents a \$137.8 billion investment, while an additional \$517.4 billion

---

[88] Australian Defence official, interview with the authors, October 2022.

[89] Australian Defence official, interview with the authors, October 2022.

[90] Australian Defence officials, interviews with the authors, October 2022. This increasing appetite to fast-track technologies and address the "valley of death" in innovation is addressed later in this chapter, when we discuss strengths and note the intent to establish an agency that focuses on the "pull-through" of innovative technologies into service.

[91] Austin Wyatt and Jai Galliott, *Toward a Trusted Autonomous Systems Offset Strategy: Examining the Options for Australia as a Middle Power*, Australian Army Occasional Paper No. 2, 2021.

[92] Eric Chewning, Will Gangware, Jess Harrington, and Dale Swartz, *How Will US Funding for Defense Technology Innovation Evolve?* McKinsey and Company, November 4, 2022.

[93] Hellyer and Stevens, 2022.

[94] Australian Academy of Science, *2023–24 Pre-Budget Submission*, January 2023.

[95] This figure relates to 2021; see Tim Brennan, Hazel Ferguson, and Ian Zhou, "Science and Research," *Budget Review 2022–23*, Australian Parliament, April 2022.

[96] Mark Boroush and Ledia Guci, *Science and Engineering Indicators 2022: Research and Development—U.S. Trends and International Comparisons*, National Science Board, April 28, 2022.

was invested by U.S. businesses.[97] Although business investment also surpasses government spending in Australian R&D investment, the university sector plays a more significant role in Australia.

Partially as a result of this lower R&D capacity, Australia has consistently leveraged overseas innovation to retain a capability offset in the region.[98] Between 2018 and 2022, the ADF was the fourth-largest arms importer globally,[99] despite ranking 13th in terms of overall military expenditure.[100] The overall result is that the ADF relies heavily on importing proven technologies (such as the K2 Huntsman Self-Propelled Howitzer), supplemented by limited domestic-international–partnered innovation efforts (most prominently the MQ-28 Ghostbat).

Budgeting and Execution

A strength of the Australian defense system is that unapproved program funds in the IIP are relatively fungible. The VCDF oversees a semiannual process (in May and December) to review and adjust the IIP based on government priorities, domestic considerations, and/or international operations.[101] IIP programs can be shifted left or right—that is, to accelerate or delay an acquisition investment—as needed. If a project is shifted left, it is accelerated at the expense of other programs. If it is shifted right, it is deprioritized so that other programs can be brought forward toward approval and funding. The IIP can be changed in its regular updates to align with new priorities, meaning that unapproved projects can be added (through a gate 0 process) or canceled.

Other types of Australian funding are also fungible in that they can be shifted across the defense portfolio, including across groups and military services. This reflects short-term flexibility in resource management, albeit with constraints on shifting money among the four major buckets (acquisition, personnel, sustainment, and operating). Fiscal policies do not permit funding to be moved in-year between such allocations, although the same effect can be achieved through managed underspend and overspend within the buckets, so long as the Defence portfolio manages expenditure within overall allocations and accurately reflects actual expenditure in annual reports.

Funding is appropriated by mission and program. Some missions, including "operations contributing to the safety of the immediate neighbourhood," "operations supporting wider interests," "defence contribution to national support tasks in Australia," and "Defence

---

[97] John F. Sargent, Jr., *U.S. Research and Development Funding and Performance: Fact Sheet*, Congressional Research Service, R44307, September 13, 2022.

[98] Wyatt and Galliott, 2021.

[99] Pieter D. Wezeman, Justine Gadon, and Siemon T. Wezeman, "Trends in International Arms Transfers, 2022," Stockholm International Peace Research Institute fact sheet, March 2023.

[100] SIPRI, undated, data as of April 2023.

[101] Australian Defence official, interview with the authors, October 2022.

people," are broad and inherently joint in their budgetary focus.[102] Others exist at the department or program level, including specific departments and programs within Defence, as well as partnerships with other Australian institutions and government departments.[103] Funding for the top 30 capital acquisition and sustainment projects is specified in the Portfolio Budget Statement.[104]

As noted, the ODCS can be accelerated during the requirements stage by combining gates 1 and 2 if there is an urgent need or technological shift. The Australian Parliament can also appropriate additional or new funding for a program deemed immediately necessary. Furthermore, the CFO "can flex the funds" within the IIP to get necessary funding for an emergency requirement, although "it comes with hurt," in the words of one interviewee.[105]

Defence is fairly unique among Australian government departments in that the IIP is a source of funds that can be drawn on for new projects. However, approval is required before shifting funds from the IIP to operating budgets.[106] As discussed earlier, the IIP permits unapproved budget shifting on a biannual basis—even among the services—to respond to geopolitical and economic conditions. Capability managers therefore possess a reasonably large degree of flexibility in how they spend the operating funds allocated to them, but they are still responsible for achieving the outcomes to which they committed in the Portfolio Budget Statement.

In October of the year following an appropriation of defense funds, the prime minister and the cabinet submit their *Annual Performance Statement*. This document "[r]eports on the actual performance results for the year against the forecasts made in the corporate plan and Portfolio Budget Statements."[107] It additionally "[p]rovides an analysis of the factors that contributed to the entity's performance results."[108]

Government departments are given the opportunity to provide Portfolio Additional Estimates Statements that reflect budget appropriations and changes between budgets. The purpose of these statements is to inform parliament and the public about changes in outcomes since the release of a given year's budget.

The operating budget for Defence expires at the end of the financial year in which it was financed. However, major procurements are handled separately through the IIP, and these

---

[102] See Australian Department of Defence, 2022a, p. 28.

[103] See Australian Department of Defence, 2022b, pp. 36–41.

[104] Australian Department of Defence, 2022b, pp. 107–123.

[105] Australian Defence official, interview with the authors, October 2022.

[106] The approving authority depends on the complexity, risk, and value of the project, and the NSC delegates some projects to a joint approval by the defence and finance ministers. See Australian National Audit Office, *Defence's Administration of the Integrated Investment Program*, Auditor-General Report No. 7 2022–23, 2022, pp. 24–25.

[107] Australian Department of Defence, 2022b, p. ix.

[108] Australian Department of Defence, 2022b, p. ix.

individual projects do not expire. Still, the overall acquisition program is expected to hit a target annual expenditure level.

Oversight

As specified, MPs and senators read the prime minister's annual budget, including the defense budget, on budget night, which is typically the second Tuesday in May. Because the appropriations bills are debated in the House of Representatives for several weeks in May and June, a summary document titled *The Particulars of Certain Proposed Expenditure* is submitted to the relevant Senate committee—in this case, Foreign Affairs, Defence, and Trade—for review concurrent with the House debate. This review is usually completed by the end of May. The House of Representatives may make substantive changes to the budget, including the defense budget, before it is passed to the Senate for final approval. There, further negotiations may be required to reach a vote in the government's favor. (The Senate cannot modify the appropriations bill.) The Senate committees review supplementary funding requests in October.[109]

Thus, the legislative review process occurs as part of the budgetary process. There are in-year forecasting and reporting mechanisms—notably, the Portfolio Additional Estimates Statements.[110] There is no legislative review of individual budget items after they have been approved as part of the budget process, other than through the biannual Senate Estimates process.[111]

The executive branch of the government holds the purse strings. Therefore, the legislature does not impose official guidance on the executive beyond ordinary legislative accountability requirements. The formal capacity of individual MPs and senators to influence cabinet decisions is limited to public scrutiny and asking questions. However, the nature of the parliamentary system means that they can occasionally affect decisions through political leverage, such as through lobbying cabinet members. Ultimately, though, the cabinet has the final say.

The DCAP process, which we described earlier, typically operates on a continual two-year cycle, which precedes the typical two-year cycles leading to gates 1 and 2. Therefore, proposal decisions must be made at least six years in advance of their inclusion in a draft Portfolio Budget Statement. (As noted, an exception would be when especially important programs are accelerated somewhat by combining gates 1 and 2 for review by relevant committees.)

Several accountability mechanisms are built into the finance regulations within Defence and the public service more broadly. These policy mechanisms include the 1988 Charter of Budget Honesty and the 2021 Charter of Budget Honesty Policy Costing Guidelines.

---

[109] Australian Senate, 2023.

[110] As with the Portfolio Budget Statement, the purpose of Portfolio Additional Estimates Statements is to inform parliament and the public about government spending and any changes since the budget was announced six months prior. The statements also provide information on new projects and their impacts on government financial and nonfinancial performance.

[111] Senate Estimates is a formal process whereby Senate committees are given the opportunity to question government department representatives about financial and nonfinancial performance against outcomes.

Defence tracks expenditures in accordance with the Defence Finance Policy Framework to ensure that what is spent matches what has been approved by the legislature. The Portfolio Additional Estimates Statements and midyear financial outlook help ensure accountability to the public. Finally, the Australia National Audit Office conducts independent government audits in a manner similar to that of the U.S. Government Accountability Office and the UK National Audit Office. Defence recently introduced in-year forecasting to better control the budget and prevent overspending, albeit with initial problems, as previously noted.

In October of the year following the disbursement of funds, Defence releases its *Annual Performance Statement*, reporting "on the actual performance [estimates for the previous financial] year against the forecasts made in the corporate plan and Portfolio Budget Statements." The statement also "provides an analysis of the factors that contributed to the entity's performance results."[112] Defence's Contestability Division "contributes [throughout the lifecycle process] to the Integrated Force Design [Process] to ensure that decisions are based on robust information, unbiased analysis process and aligned with strategic priorities."[113] The Australian National Audit Office also executes audits in an independent manner.

Compared with DoD, Defence receives significantly less PPBE guidance from the Australian legislature. The executive branch—the Minister for Defence, the prime minister, and cabinet colleagues—hold the purse strings. The other MPs and senators can review Defence's PPBE-like functions and direct their questions to either the Minister for Defence or directly to Defence through parliamentary liaison officers.

The Defence Strategic Review

In August 2022, the Australian government initiated a substantial review to assess whether Australia had the necessary defense capability, posture, and preparedness, given the strategic circumstances. The unclassified report of the DSR was publicly released on April 24, 2023, and addressed aspects of Defence's PPBE-like processes.[114]

A key finding of the DSR is that Australia's force structure was based on a balanced force model that no longer reflected the strategic environment. It recommends that new capabilities be developed to reflect a strategy of deterrence to deny an adversary freedom of action to militarily coerce Australia and to operate against Australia without being held at risk. The review highlights the importance of translating this policy into a force structure that can be realized within required time frames and resources. Force structure goals are grouped in three temporal tranches of enhancing the force in being (2023–2025), accelerating toward the objective integrated force (2026–2030), and delivering the future integrated force (2031 and beyond). Importantly, the review identifies that, in order to pursue high-priority capa-

---

[112] Australian Department of Defence, 2022b, p. ix.

[113] Australian Department of Defence, 2021a, p. 34.

[114] Australian Department of Defence, *National Defence: Defence Strategic Review*, 2023.

bilities to achieve this force structure, lower-priority programs would need to be canceled or delayed.[115]

The DSR notes that defense planning is a matter of managing strategic risk and that defense spending must reflect the strategic challenges that the country faces. Accordingly, the DSR notes that funding would need to be increased to meet these circumstances. The full cost of implementing the review cannot be quantified until the recommendations can be costed. Accordingly, the defense budget within the forward estimates has not changed.

Chapter 12 of the DSR notes that capability acquisition processes in Australia need to be made more efficient. It also notes that capability managers have too much latitude to specify design changes that complicate acquisition. The review recommends that processes should focus on minimum viable capability and required time frames, with more sole-source and off-the-shelf procurement and fewer design modifications. Processes for urgent and strategically important projects should be streamlined.

Chapter 9 of the DSR also notes the erosion of technological superiority and that Australia should consider options to achieve asymmetric advantage. It also highlights the importance of the second pillar of the AUKUS agreement for advanced technologies, including the contribution of innovation through the Defence Science and Technology Group's oversight of innovation. Such innovation needs to be encouraged, which will be supported by the Australian Strategic Capabilities Accelerator (ASCA) initiative.

Australian Strategic Capabilities Accelerator

The ASCA was launched by the Australian government shortly after the DSR in April 2023.[116] The ASCA is focused on supporting and assessing innovative defense solutions at relatively high TRL, where progression through acquisition into service has had limited success in the past. The ASCA will use governance arrangements to ensure that truly innovative systems can be introduced into service to enhance defense capabilities. The ASCA will supersede and expand on Australia's extant defense innovation processes and industry engagement, such as the Next Generation Technologies Fund and the DIH.

Analysis of Australia's Defense Budgeting Process

Strengths

Ministerial oversight of Defence and its annual budget theoretically ensures the responsible use of resources. Four-year forward estimates and ten-year baseline budget approvals,

---

[115] Australian Department of Defence, 2023.

[116] The ASCA was launched by the Minister for Defence on April 28, 2023. See Richard Marles and Pat Conroy, "Government Announces Most Significant Reshaping of Defence Innovation in Decades to Boost National Security," press release, April 28, 2023.

informed by strategic documents, provide a measure of consistency for both ministerial oversight and the capability managers overseeing programs. In sum, as several interviewees explained, the defense budget is largely already set.[117] The introduction of the ODCS in 2015 better centralized Defence's PPBE system, providing greater oversight of programs and tying plans to resources. This centralized planning approach and the analytic support of such processes as the DCAP have meant that legacy capabilities and replacement philosophies have been more effectively scrutinized and resource allocations are more tightly linked to overall defense priorities. Execution is delegated to capability managers, who, in turn, delegate responsibilities to lead delivery groups for a more effective and efficient process.[118] The Australian National Audit Office provides accountability, as does the Portfolio Budget Statement, given that it is a public document that explicitly links resources to government outcomes.

The contestability function informs oversight but is not oversight *itself*. Rather, contestability advice is integrated into and informs the decisionmaking of the Defence Investment Committee, the Defence Finance and Resources Committee, and the NSC. Oversight also exists through independent reviews of acquisition activities and through Senate reviews of defense programs. The Portfolio Budget Statement is subject to parliamentary and public scrutiny. Although the opposition can rarely change the government's defense decisions, grievances can be aired in public, thereby pressuring the government as elections loom.

Recently, there have been cumulative defense acquisition budget and procurement schedule overruns.[119] Concerns about these overruns may be overblown, however. Marcus Hellyer of the Australian Strategic Policy Institute explained that budget "blowouts" often are actually the result of either a deliberately "staged increase in capacity" or "fluctuations in exchange rates."[120] Additionally, some programs have been reported to be late because, even though the capabilities are in service, the project has not been closed because of minor outstanding elements. Hellyer has concluded that Defence "develops second-pass cost estimates extremely conservatively, putting risk margins on top of risk margins so that there's virtually no prospect of going over budget. But that can tie up funds that could be used for other priorities."[121]

The Portfolio Budget Statement and the IIP generally reflect the value to the warfighter of resource allocations and expenditures, given that they are derived from mission needs and strategic priorities. The DCAP ensures that the ADF's current and planned force structure is fit for purpose against prospective operational scenarios, taking into account theater campaign plans, operational concepts, and preparedness directives.

The *Defence Capability Manual* prioritizes the need for value for money at every stage of the defense budgeting process. "Value for money," it states, "should not be seen simply as

---

[117] Australian Defence officials, interviews with the authors, October 2022.

[118] Australian Department of Defence, 2021a, p. 5.

[119] "Australian Defence Projects Billions over Budget, Decades Late," Reuters, October 9, 2022.

[120] Hellyer, 2022.

[121] Hellyer, 2022.

economising or as a purely financial concept. A cheaper option will still need to be fit for purpose and offer real value. On the other hand, a more expensive option needs to be commensurably better than a cheaper option to be preferred."[122] Projects are expected to finish on time and on budget to maximize value for the warfighter. Overspending is dangerous for programs because it results in a deficit that Defence needs to cover; underspending can lead to suspicion of underperformance or "underachievement."[123]

Strategic plans are linked to budgets, first through the IIP resulting from the DCAP process, and then through the formal budget request. The contestability function ensures "that the requirements and the resultant capabilities delivered to the [ADF] are aligned with articulated strategy and agreed-upon resources."[124]

As part of the annual budgetary process, Defence is provided with assurance of sustained funding levels over a four-year rolling period, including the current fiscal year (the forward estimates period). The *2016 Defence White Paper* laid out an indicative baseline for defense spending (except operating costs) over ten years. Sounding the alarm about Australia no longer being able to rely on a ten-year warning time for a major conflict, the *2020 Defence Strategic Update* laid out an updated version of this baseline, extending it to 2029–2030. Furthermore, Defence plans its investments out as far as 20 years as whole-of-life investments,[125] incorporating the IIP, the *Force Structure Plan*, and the classified *Defence Strategic Workforce Plan*, along with "other plans intended to coordinate the delivery of specific FIC[s] and plans for further capability analysis, innovation, and research and development related to new potential options."[126]

Parliament debates the defense budget and explanatory Portfolio Budget Statement on an annual basis as part of the overall budget approval process. However, the government can boost the defense budget in periods of national emergency (e.g., wildfires) or overseas military operations (e.g., Iraq, Timor-Leste) using the no-win/no-loss model.[127] Similarly, the government can supplement the department's allocation to alleviate inflationary pressures beyond those forecasted. At any time, the NSC can consider urgent priorities and their funding implications, and the Minister for Defence can intervene to prioritize certain programs or investments. There is flexibility to move current-year funds among groups and military services to meet emerging or unforeseen needs. The CFO can also divert funding to meet emerging priorities. The small size of the ADF facilitates flexibility and jointness but at the

---

[122] Australian Department of Defence, 2021a, p. 3.

[123] Australian Defence official, interview with the authors, October 2022.

[124] Cook et al., 2016, p. iii.

[125] Australian Defence officials, interviews with the authors, October and November 2022.

[126] Australian Department of Defence, 2021a, p. 32.

[127] No-win/no-loss funding is appropriated through appropriations bills. It can be appropriated to offset the cost of approved operations and foreign exchange movements.

potential loss of enduring capability.[128] The DCAP is the key activity that promotes agility, but it is linked to government updates of strategic guidance, which may not be sufficiently agile.[129] However, there has been an effort to make these updates more frequent and ongoing.[130] The IIP has moved beyond mere platform replacement, and in the future, capability managers could undertake relevant analysis to further increase agility.

A view that was shared during our interviews was that Australia's defense budgeting process is transitioning from one "epoch" to another that will be more agile and flexible.[131] Interviewees considered current processes to be effective but "turgid." In the new epoch, there could be agreement to trade some of the openness, transparency, and risk aversion of the current processes to increase responsiveness and agility. This would mean a deliberate change in processes, policies, and culture to allow the rapid introduction of capabilities. Another possibility for increasing agility would be to extend the no-win/no-loss provision for operations to ordering ordnance and other expendables prior to a conflict, so that the ADF would be more thoroughly and rapidly prepared for emerging threats. Recently, stakeholders have noted a shift in the DCAP/IIP processes from deliberate to agile, focusing risk reduction on operations rather than acquisitions. The proposed ASCA would emphasize the "pull-through" of innovations into service. This would require that the ASCA develop the agility to move funds more effectively between projects, as well as more efficiently allocate them in a timely and effective manner.

Challenges

Although the introduction of the ODCS in 2015 centralized Defence's PPBE system, it did not necessarily accelerate Australia's ability to respond to emerging and rapidly evolving threats with speed and agility. The nominal timeline for the requirements process in the ODCS is two years per pass, or four years from gate 0 to gate 2, although it is often longer. It often takes up to a decade for a new acquisition to go from conception to being funded in the IIP. A combined pass can somewhat reduce this time frame. Nominally approved programs can remain unfunded in the IIP for a long period, owing to a variety of geopolitical, domestic, or economic factors. Such programs may also simply not be competitive priorities. Public awareness of approved but unfunded programs can lead to industry expectation of work and potential inertia to change. Even after programs are funded, there is no guarantee that they will be completed on time or at cost. Ironically, projects may appear to run excessively long and overbudget because conservative budgeting—driven by concerns about cost, schedule, and risk—ultimately constrain efforts to be more agile. In general, a legacy of cultural aversion to risk constrains the ADF's budgeting and programming flexibility.

---

[128] Australian Defence official, interview with the authors, November 2022.

[129] Australian Defence official, interview with the authors, October 2022.

[130] Australian Defence official, interview with the authors, October 2022.

[131] Australian Defence official, interviews with the authors, October and November 2022.

Other issues slow down Australia's defense budgeting process. Large amounts of funding are fixed in contract commitments, constraining the agility to meet emerging needs. Acquisition processes that are transparent and promote competition can make it difficult to undertake quick, sole-source arrangements. Much of Australia's capability is of U.S. origin, which offers the benefit of interoperability, but it means that Australia either must wait for U.S. systems to be developed or is forced to introduce capabilities that are not interoperable. This problem is exacerbated by a lack of allied engagement in requirements processes. Accelerating DoD agility therefore would benefit Australia (and other allies).[132] Australia's small defense industry can contribute to niche capabilities but is less competitive in developing complete systems. There are constraints, as we noted earlier, on the rapid acquisition of innovative solutions. Defence must compete with other government departments for finite resources.

Australia also suffers from workforce challenges; it has an insufficient pool of qualified personnel to join Defence and the defense industry.[133] COVID-19, "supply chain [cost] increases," employment levels, and general economic conditions are factors in this workforce deficit.[134]

Finally, Defence does not support truly novel or innovative solutions; the DIH prioritizes extant plans and technologies.[135] Australian technologies continue to suffer from the valley of death. Limited Australian sovereign industry is logically directed toward sustainment—more so than innovation. Most Australian R&D funding has been focused on low-TRL systems, and limited attention has been paid to pulling through more-mature developments, although the aforementioned establishment of the ASCA could improve this situation.

Applicability

Australia's defense budgeting system is similar to that of the United States and the UK in several important respects. It is guided by a series of strategic planning documents that are updated every four years or so and lays out major capital projects, resources, and goals. As in the UK, strategic documents provide a baseline for the defense budget for a coming period—in this case, ten years—which allows for a relatively high degree of budget certainty. The IIP approach provides accountability and structure for project development while also offering flexibility. Like DoD, Defence possesses technology facilitators, such as DIH, that help inte-

---

[132]   Australian Defence official, interview with the authors, October 2022.

[133] This workforce deficiency is evident within the department. See James Massola, "Defence Facing a 'Personnel Crisis' with Thousands More Uniformed Members Needed," *Sydney Morning Herald*, November 14, 2022. On the military support industry, see Leah MacLennan, "Warning of Defence Shipbuilding Skills Shortage Amid Uncertainty over Local Submarine Build," Australian Broadcasting Corporation News, May 10, 2022.

[134] Australian Defence officials, interviews with the authors, October 2022.

[135] Gregor Ferguson, "Peever Review Could Transform Defence Innovation and Acquisition," *The Australian*, October 30, 2021.

grate emerging technologies with defense priorities in a coherent, organized fashion. There are still some criticisms of DIH's effectiveness, and there are few examples of the successful adoption of new innovations through that program. Concurrent with our interviews, Defence was discussing the introduction of the ASCA to help fast-track innovations into service. However, some observers have acknowledged that that agency's success will be highly dependent on broader changes to PPBE-like processes to facilitate agility.

Although it is rare, the Minister for Defence can personally promote programs of government priority even if the department disagrees, potentially accelerating new technologies and priorities. Defence has access to independent auditors, such as the Australian National Audit Office, which is similar to the U.S. Government Accountability Office and UK National Audit Office. Other key advantages include the lack of continuing resolutions in the legislature and the capability program system, which is designed to develop projects holistically and is akin to the U.S. DOTMLPF framework.

Finally, and arguably most importantly, the government can shift unapproved programs in the IIP biannually, including between the services—an option that is generally not available to U.S. defense leaders. Unlike in the United States, Defence's decisions can be made without direct legislative oversight.[136]

One notable feature of the ADF is that it operates in a relatively more joint manner than its U.S. counterpart. Although this is easier given the ADF's smaller size and distinct organization, Defence has not always been so integrated. The department comprised separate governance structures for each service until the late 20th century and early 21st century, as reforms led to a greater emphasis on joint planning and budgeting authorities and processes. As a result of these efforts, the modern ADF operates in a comparatively joint manner; for example, some program costs, such as shared fuel costs, are centralized.[137] There is also a level of joint financial governance, with the service component CFOs reporting to the departmental CFO and to their service chiefs. These points may be important to the U.S. defense community, given ongoing efforts to enhance jointness across the U.S. military.

Australia's small defense industry allows a greater ability to pivot, albeit with caution, to maintain consistency of the force's employability against a variety of threats and in a variety of missions. Centralized governance enables fungibility to move in-year funds across Defence to meet changing priorities. There is a shift occurring in risk focus from acquisition processes to operational capability, and this has likely streamlined the ODCS process for certain urgent capabilities while enhancing flexibility.

However, there are challenges too. As with DoD's PPBE System, it may take years for a proposed capability to progress through the ODCS process before acquisition begins. Some

---

[136] Australian Defence official, interview with the authors, November 2022. Parliament has the ability to review many decisions through question time and Senate Estimates hearings, but most decisions (other than appropriations) are made by the executive (i.e., prime minister, cabinet, Minister for Defence).

[137] Australian Defence official, interview with the authors, November 2022.

programs can remain unapproved in the IIP for years. There is no guarantee that funded capability programs will be completed on time and at cost.

There are relatively few modes to accelerate innovative programs or respond rapidly to emerging threats outside DIH and the Defense Science and Technology Group. Defence can combine gates 1 and 2 to somewhat accelerate the process, and the Minister for Defence can intervene, as was done with the acquisition of F/A-18 *Super Hornets.* In light of the AUKUS agreement, the question of how Australia can truly accelerate its defense spending and procurement process remains. Australia likely will need to expand its use of expeditious processes in cases in which operational urgency necessitates exchanging risk aversion for agility.

Interviewees discussed this shift in terms of moving from a focus on mitigating acquisition risks to mitigating operational risks, in recognition of the deteriorating strategic environment. Additionally, interviewees discussed the importance of keeping pace with threats. While doing so may result in more-streamlined processes for new capabilities, there is value in the deliberate plans and certainty of the current process. Thus, there may be elements of the extant ODCS that need to be retained, and the balance between deliberate and agile may depend on the urgency of individual capability proposals. This two-speed approach might build on previous processes used by Defence to expedite urgent operational requirements.

Lessons from Australia's Defense Budgeting Process

Deliberate PPBE processes give Defence more certainty and greater alignment of strategy and resources, but, if applied across all requirements, they can impede agility. Centralization of priority-setting and resource decisionmaking is an advantage in the Australian system. The U.S. relationship is key to the agility (or lack thereof) of Australian processes; enhanced cooperation may be of mutual benefit not only in technology but also upstream in PPBE processes.

Lesson 1: Deliberate Processes

The defense planning, programming, and budgeting processes used in Australia provide a high level of certainty for the development and operationalization of military capabilities. The processes ensure a strong connection between strategy and resources, reduce prospects for the misuse of funds or inefficiency, and pose a limited risk of blocked funding from year to year. The deliberate processes can be an impediment to agility, however, if they are universally applied. Nonetheless, Defence has mechanisms to support more-agile changes in capabilities and is committed to fast-tracking innovative systems.

Lesson 2: Centralization

The centralized approach adopted by Defence reduces the authority of individual military services, promotes jointness, and prioritizes the best capabilities to achieve effects rather than to reinforce legacies. Whereas the services retain accountability as capability managers of their respective resources, priorities and resources can be adjusted to account for emerging threats and opportunities, even within the current financial year. This flexibility is facilitated largely by (1) the organizational arrangements that make the central committees (Defence's internal Investment Committee and Defence Committee) responsible for overall resource priorities and (2) the organizational structure that makes finance managers accountable both to functional groups and to the CFO.

Lesson 3: U.S. Relationship

Defence has a significant dependence on DoD. This dependence arises from the capacity of the U.S. industrial base and the technological edge of its systems, but it also arises from the high priority that Australia places on allied interoperability. In a similar vein, both the United States and Australia have demonstrated an interest in some areas in the interchangeability of equipment and processes, enabling options to use one another's systems. Interchangeability is becoming more relevant in such applications as common maintenance and resupply of ordnance. Typically, Defence is a customer of U.S. systems, often through the FMS process. Because the development and production of these systems may depend, in turn, on U.S. PPBE processes, there are limitations to Australia's ability to become more agile than those U.S. processes will allow—at least with respect to major weapon systems. This constraint is acceptable to Defence, in view of the interoperability and capability advantages. During discussions about AUKUS, emerging technologies, innovation, and weapon cooperation, our interviewees indicated that the U.S.-Australia relationship may shift to one in which Australia is not simply a defense materiel customer but more of a partner. Beyond technology cooperation, there is the prospect that greater transparency and coordination across DoD's and Defence's PPBE processes could lead to mutual benefits in terms of capability agility, synergies, and efficiencies. The new ASCA, in conjunction with the AUKUS initiatives, could facilitate allied coordination and the rapid introduction of new capabilities into service.

Each of these lessons cuts across all phases of planning, programming, budgeting, execution, and oversight. In Table 2.2, key features of the lessons are organized into four themes: decisionmakers and stakeholders, planning and programming, budgeting and execution, and oversight.

TABLE 2.2

Features of Australia's Defense Budgeting Process

Theme	Features	Description
Decisionmakers and stakeholders	Centralized within government, with a parliamentary system.	The government, under the party in power, holds the purse strings.
Planning and programming	Planning and programming are centralized through the DCAP.	The DCAP converts strategy into tangible capability programs.
	Defence is dependent to a great degree on DoD.	Dependence arises from the capacity of the U.S. industrial base and the technological edge of its systems, along with allied interoperability.
Budgeting and execution	The IIP ensures that capability programs are properly budgeted.	The IIP informs the Portfolio Budget Statement that, in turn, funds the programs.
	Capability managers are responsible for both sustainment (execution) and disposal.	The same capability manager is responsible during and after acquisition.
Oversight	The Australian National Audit Office is similar to the U.S. Government Accountability Office and UK National Audit Office.	The Australian National Audit Office provides independent oversight and assessment.
	There is a culture of risk aversion in the Australian government.	Stakeholders seek to spend within limits, potentially limiting agility while adhering to the annual budget.

CHAPTER 3

Canada

Devon Hill and Yuliya Shokh

Canada and the United States have a long, collaborative defense relationship. Their militaries have fought alongside one another in several conflicts since World War II. Canada describes the United States as its "most important ally and defence partner,"[1] and the U.S. Department of State says that the two countries' "bilateral relationship is one of the closest and most extensive."[2] Both Canada and the United States are members of NATO, and they cooperate extensively through multiple bilateral defense forums and agreements, including NORAD, the Permanent Joint Board on Defense, the Military Cooperation Committee, the Combined Defense Plan, the Tri-Command Framework, the Canada-U.S. Civil Assistance Plan, the Canada-U.S. Civil Assistance Plan, and the National Technology Industrial Base.[3]

But the United States and Canada have different approaches toward defense spending. The United States and Canada spend vastly different amounts on defense annually: The United States appropriated about U.S. $798 billion in FY 2023, and Canada spent roughly one-40th of that sum, or U.S. $19 billion, in FY 2022–2023.[4] Canada's parliamentary system also operates much differently than the U.S. political system. Nonetheless, a review of Canada's defense budgeting process can provide U.S. policymakers with useful insights on resource allocation methods and challenges.

For example, a key strength is that Canada recognizes its status as a middle power and has historically sought to increase its relative influence through multilateral diplomacy and contributions to alliances;[5] this alliance-oriented foreign and defense policy approach has helped offset relative personnel and resource limitations. Furthermore, the DND has sought to improve budget transparency in recent years, allowing for better long-term spending pro-

---

[1] Government of Canada, 2014.

[2] U.S. Department of State, 2022.

[3] Government of Canada, 2014.

[4] U.S. House of Representatives, Committee on Appropriations, undated; Government of Canada, 2022b. The Canadian government's fiscal year runs from April 1 of the current year to March 30 of the following year. For example, FY 2022–2023 began on April 1, 2022, and ended on March 30, 2023.

[5] Meyer and Fergusson, 2021.

jections, although challenges remain that we discuss in further detail later in this chapter. The DND has also taken a service-agnostic acquisition process that weighs new projects against strategic priorities.

There are some challenges that provide context for some of the lessons that close out this chapter. There appears to be little political appetite for defense spending growth in Canada, which limits its ability to quickly reach NATO's goal of spending 2 percent of GDP on defense, for example. There is also limited bureaucratic capacity to absorb that new spending quickly.

This chapter sheds light on how the DND develops high-level strategies, prioritizes resource allocation, works within Canada's federal budgeting process, and executes its missions. After an overview of Canada's government, including DND and its defense policy statements, we describe Canada's federal government budgeting process, noting the nonexistence of funding lapses and the widespread use of planning and accountability reports. We then review Canada's defense budgeting process, including its accounting practices, recent defense spending levels, important decisionmakers and stakeholders, planning and programming, budgeting and execution, and legislative scrutiny and oversight. We conclude with a summary of Canada's budgeting strengths and challenges, followed by lessons learned.

Overview of Canada's Government

Canada has a mixed system of government: It is a federation of provinces and territories, a constitutional monarchy, and a parliamentary democracy. It has a strong central federal government led by the parliament, which shares domestic policy responsibilities with the governments of the country's ten provinces and three territories. Figure 3.1 details Canada's federal executive, legislative, and judicial institutions. In simple terms, the formal head of the Canadian state is the British monarch, represented by an appointed governor general in whom executive and legislative power is vested. In practice, the lower chamber of parliament, the House of Commons, is elected by voters; MPs represent one of 338 federal electoral districts (often called *ridings* in Canada). The leader of the largest party in the House of Commons becomes prime minister and selects the cabinet. The upper chamber, the Senate, consists of 105 members appointed by the governor general on the recommendation of the prime minister. Although the Senate technically has formidable legislative powers, it rarely exercises them in practice. For example, the Senate may block legislation or insist on amendments that have been rejected by the House but will typically make only clarifying or simplifying

FIGURE 3.1

Canada's System of Government

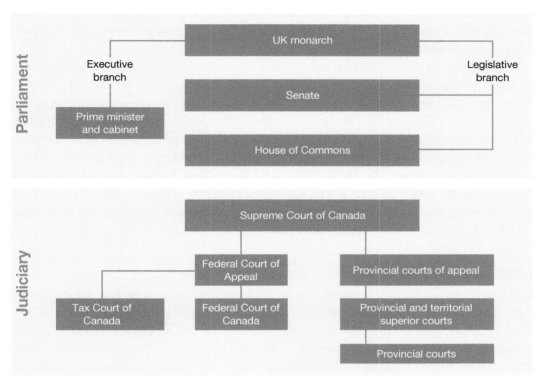

SOURCE: Adapted from Forsey, 2020, p. 32.

amendments that are almost always accepted.[6] Defense, as a matter of national interest, is an exclusive power of the parliament and thus the federal government.[7]

As the head of the largest party in the House of Commons, the prime minister holds significant power. If the prime minister's party commands a majority of members of the House

---

[6] Eugene A. Forsey, *How Canadians Govern Themselves*, 10th ed., Library of Parliament, March 2020, pp. 32–36. In simple terms, senators are distributed by region rather than by province or population. Twenty-four senators each represent the Maritime provinces (ten each from Nova Scotia and New Brunswick and four from Prince Edward Island), Quebec, Ontario, the Western provinces (six each from Manitoba, Saskatchewan, Alberta, and British Columbia), six from Newfoundland and Labrador, and one each from Yukon, Northwest Territories, and Nunavut. Senators must be at least 30 years old and may hold office until the age of 75. Since 2016, potential candidates have been vetted by the Independent Advisory Board for Senator Appointments; until that point, senatorial appointments were largely based on partisan affiliation, but all appointees since have been unaffiliated. Since taking office in November 2014, Prime Minister Justin Trudeau has appointed 66 senators.

[7] Government of Canada, Justice Laws Website, "The Constitution Acts, 1867 to 1982: Section VI, Distribution of Legislative Powers, Powers of the Parliament," webpage, updated February 17, 2023.

of Commons, the prime minister's power is generally limited only by the political party they lead, external political pressures, or elections. If the prime minister leads the largest party but that party does *not* hold a majority of seats in the House of Commons, the prime minister may still form a cabinet and is in a strong position to pass legislation but relies on other parties for the passage of such legislation.

Prime Minister Justin Trudeau, the leader of the federal Liberal Party of Canada, led a majority government from 2015 to 2019 and, as of this writing, has led minority governments since 2019. In early 2022, he signed a confidence-and-supply agreement with a smaller, left-wing party called the New Democratic Party.[8] The agreement allows the Liberal Party to govern until the next election in 2025, contingent on the implementation of a negotiated list of policies; in return, the New Democratic Party agreed to support confidence measures, including budget and spending bills. Without this agreement, Trudeau's government would have been at risk of failing; if a prime minister and cabinet are defeated in the House of Commons on a spending or confidence vote, they are required to resign, and, typically, an election for a new government is held shortly thereafter.[9]

DND Overview

DND, as its name suggests, is responsible for implementing policy regarding the defense of Canadian interests at home and abroad. At any time, the Canadian Armed Forces (CAF) need to be prepared to undertake missions to protect the country and its citizens and to maintain international peace and stability.[10] According to DND's 2017 defense policy statement, *Strong, Secure, Engaged: Canada's Defence Policy*, there are three primary components to Canada's defense vision:

- **Strong at home**, its sovereignty well-defended by a Canadian Armed Forces also ready to assist in times of natural disaster, other emergencies, and search and rescue;
- **Secure in North America**, active in a renewed defence partnership in NORAD with the United States;
- **Engaged in the world**, with the Canadian Armed Forces doing its part in Canada's contributions to a more stable, peaceful world, including through peace support operations and peacekeeping.[11]

---

[8] Aaron Wherry, Rosemary Barton, David Cochrane, and Vassy Kapelos, "How the Liberals and New Democrats Made a Deal to Preserve the Minority Government," CBC News, March 27, 2022.

[9] Forsey, 2020, p. 4.

[10] DND, "Mandate of National Defence and the Canadian Armed Forces," webpage, updated September 24, 2018d.

[11] DND, *Strong, Secure, Engaged: Canada's Defence Policy*, 2017, p. 14.

The 2017 policy document highlights what DND will do to "succeed in an unpredictable and complex security environment," including the following:

- Actively address threats abroad for stability at home;
- Field an agile, well-educated, flexible, diverse, combat-ready military; . . .
- Act as a responsible, value-added partner with NORAD, NATO and Five-Eyes partners;
- Work with the United States to ensure that NORAD is modernized to meet existing and future challenges;
- Balance traditional relationships with the need to engage emerging powers. . . .[12]

The policy lays out core CAF missions, including detecting, deterring, and defending against threats to or attacks on Canada or North America. According to the policy document, CAF will be prepared to sustain several operations concurrently, including meeting alliance commitments. *Strong, Secure, Engaged* further establishes that CAF will defend Canada, respond domestically in support of civilian authorities, meet NORAD and NATO commitments, and contribute to international peace and stability through the following concurrent deployments:

- Two sustained deployments of ~500–1500 personnel, including one as a lead nation;
- One time-limited deployment of ~500–1500 personnel (6–9 months duration);
- Two sustained deployments of ~100–500 personnel;
- Two time-limited deployments (6–9 months) of ~100–500 personnel;
- One Disaster Assistance Response Team (DART) deployment, with scalable additional support; and
- One Non-Combatant Evacuation Operation, with scaleable additional support.[13]

DND Defense Policy Statements

Defense policy statements, such as the 2017 document, sometimes known as *white papers*, are not required to be issued on any regular basis, nor is there any standard practice for how reviews of the statements are conducted. Furthermore, such statements have no legal standing and do not authorize expenditures.[14] Since the mid-1960s, nine defense white papers have been published: in 1964, 1971, 1987, 1992, 1994, 2004, 2005, 2008, and 2017.[15] Outside observers disagree on which documents can be cited as national defense policy. One author, Eugene

---

[12] DND, 2017, p. 14.

[13] DND, 2017, p. 17.

[14] Ariel Shapiro and Anne-Marie Therrien-Tremblay, "Canada's Defence Policy Statements: Change and Continuity," HillNotes, Library of Parliament, September 22, 2022.

[15] Shapiro and Therrien-Tremblay, 2022. In 2005, the Canadian Department of Foreign Affairs, Trade and Development (now better known as Global Affairs Canada) released a five-volume series of international policy statements, one of which addressed defense policy. See Bert Chapman, "The Geopolitics of Canadian

Lang, describes the short-term relevance of white papers and their often ambitious agendas.[16] Specifically, he highlights three forces that have undermined these defense policy statements: (1) structural deficits leading to departmental spending restraint, (2) unforeseen changes in the international security environment, and (3) elections that bring new parties with different priorities to power.[17] Although these forces are not unique to Canada, the first helps explain the country's relative restraint in defense spending compared with that of other NATO countries and the United States. The second force is a timely subject, given that DND plans to issue an update to its 2017 paper, which some experts feel has been overtaken by recent events, especially Russia's invasion of Ukraine.[18] Although no date has been given for the update, it is expected to be published before the 2025 federal election—in time for the current administration to articulate its priorities before a potential change in government.

Overview of Canada's Federal Budgeting Process

A process called the Expenditure Management System guides Canada's federal budget process. The system provides the structure for budgeting and resource allocation, outlines the stages and key players, allows for public consultation and participation, charts the timeline, and lists the reporting and accountability requirements.[19] Figure 3.2 outlines the steps in the Expenditure Management System.

The Canadian government's fiscal years run from April 1 to March 31. Between March and June of each year, departments prepare and review their business plans. By June, they release their outlooks for the coming year. The budget preparation process continues more formally when the cabinet convenes sometime around June to consider the broader elements of the next budget, taking into account the economic and political climate, reports on public issues, and the government's priorities. These cabinet discussions result in plans that guide how civil servants in the Privy Council Office, Department of Finance, and Treasury Board Secretariat work with other federal departments to develop budget strategies and options for the Minister of Finance to consider.[20] In particular, the Treasury Board and Department of

---

Defense White Papers: Lofty Rhetoric and Limited Results," *Geopolitics, History, and International Relations*, Vol. 11, No. 1, 2019, p. 20.

[16] Eugene Lang, "The Shelf Life of Defence White Papers," *Policy Options*, June 23, 2017.

[17] Lang, 2017.

[18] Government of Canada, "Chapter 5: Canada's Leadership in the World," in *Budget 2022: A Plan to Grow Our Economy and Make Life More Affordable*, 2022a.

[19] Amelita A. Armit, "An Overview of the Canadian Budget Process," paper presented to a Roundtable on State Financial Control at the Ninth St. Petersburg International Economic Forum, Moscow, Russia, June 2005, p. 2.

[20] Armit, 2005, p. 3.

FIGURE 3.2

Canada's Expenditure Management System

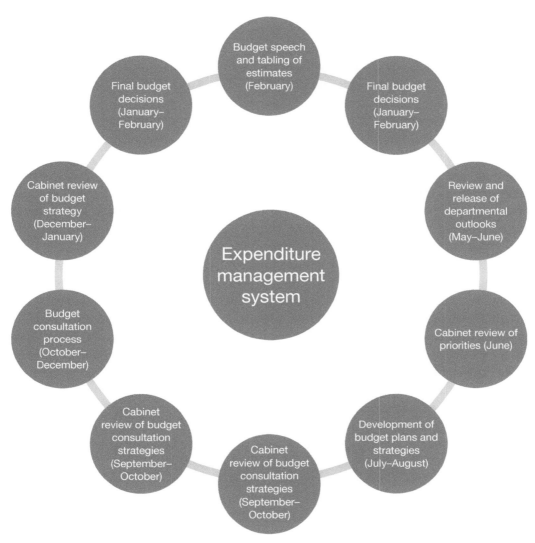

SOURCE: Adapted from Armit, 2005, p. 5.

Finance set annual spending limits for federal agencies, including DND. These limits apply to capital expenditures and determine the number of new projects funded.

By September, the consultation process begins. The Department of Finance prepares consultation papers on the country's fiscal and economic outlook, as well as prospective fiscal and spending targets. In October, the Minister of Finance releases the papers and begins consulting with the House of Commons Standing Committee on Finance, provincial finance ministers, the public, and other stakeholders. Around this time, the Minister of Finance will also give an economic and fiscal policy statement to the House, updating members on the state of the economy and any spending changes. Typically, by December, the Minister of Finance develops a budget strategy based on this consultation. In January of the year the budget is due to be presented to parliament, the cabinet reviews the budget strategy, fiscal targets, and new spending initiatives or proposed reductions. The Minister of Finance, with the prime minister, makes final decisions before the Department of Finance finalizes the budget documents. The Treasury Board Secretariat prepares the Main Estimates, which incorporate the budget decisions and detail spending by agency and department.

Usually in February, the Minister of Finance gives an official budget speech in the House of Commons. Shortly thereafter, the president of the Treasury Board presents the Main Estimates to the House on behalf of all government departments and agencies for legislative consideration. Following the presentation of the estimates, which must occur by March 1 (at the same time the next year's budget is starting to be prepared), the House begins its deliberations in various standing committees that call ministers, senior civil servants, and other interested parties to appear. This process culminates by May 31 in committee reports on those estimates.[21] By June, parliament typically approves the budget and Main Estimates.

The executive branch plays the dominant role in budget preparation, and parliament has relatively limited influence. Parliament does perform legislative and oversight functions through its review and approval of the budget.[22] When the executive controls a majority of seats in the House of Commons, it is in a very strong position to have its prepared budget approved with no or minimal changes. When the executive controls a plurality of seats but not a majority, as was the case in 2023, it relies on support from the opposition or other, smaller parties to pass budgets and other key legislation. The party that leads the government might make official or unofficial confidence-and-supply agreements that outline the legislative priorities of the participants and ensure that legislation receives a sufficient number of votes to pass.[23] If the ruling government cannot pass its budget through the House of Commons, it is considered to no longer have the confidence of the House, and an election must be called.

Funds are allocated to DND (and other federal agencies and departments) by means of supply periods, or "the process by which the government asks Parliament to appropriate funds in support of approved programs and services."[24] Each fiscal year is divided into three parliamentary supply periods during which parliament considers whether to fund Supplementary Estimates A, B, and C, respectively. Supplementary Estimates present spending that was not ready to be included in the Main Estimates. The first parliamentary supply period

---

[21] Armit, 2005, p. 3.

[22] Armit, 2005, p. 2.

[23] For example, see Justin Trudeau, "Delivering for Canadians Now," press release, March 22, 2022.

[24] Government of Canada, "The Reporting Cycle for Government Expenditures," webpage, updated June 17, 2010.

runs from April 1 to June 23, the second from June 24 to December 10, and the third from December 11 to March 26.[25]

During the first parliamentary supply period, documents for the Main Estimates and the first Supplementary Estimates (A) are tabled and voted on.[26] DND tends to receive 50–90 percent of its funding during this first period.[27] During the second period, Supplementary Estimates B and accompanying supply bills are introduced and voted on; additionally, the Public Accounts of Canada for the previous fiscal year, containing audited financial statements and departmental results reports (DRRs), are tabled in the House of Commons. The Minister of Finance also prepares an economic and fiscal update that is delivered to parliament and the public. During the third period, the Supplementary Estimates C are tabled, if needed, and the related supply bill is introduced and voted on.[28] The supply cycle is detailed in Figure 3.3.

A Process That Prevents Government Funding Lapses

Canadian government agencies are not at risk of shutdown because of funding lapses. When the federal budget has not been passed by parliament by the beginning of the fiscal year, the government can authorize continued spending at prior-year levels. If a government falls and an election is called before a budget is passed, the government requests the issuance of special warrants to secure funding to continue the normal operations of government, pay for ongoing programs, and meet contractual obligations. These special warrants are not subject to House of Commons approval but are subject to Treasury Board and cabinet approval.[29] A 2018 report noted that, although special warrants are meant to be used during election periods when a budget is not in place, they have been used on a short-term basis in other circumstances to avoid the need for a vote in parliament.[30] For instance, special warrants were used to fund government operations during the COVID-19 pandemic emergency.

Because the budget and Main Estimates are introduced so close to the beginning of the new fiscal year, an interim supply bill is typically presented and voted on to allow government funding in accordance with new spending plans. This interim supply bill, which provides funding to government departments and agencies for the first three months of the fiscal year, is typically introduced during the third supply period and voted on in March.[31]

---

[25] Government of Canada, 2010.

[26] In contrast to the U.S. usage of *tabled* (to postpone for later consideration), the term in Canadian government parlance has the opposite meaning: to bring an issue forward for debate.

[27] Canada subject-matter expert, interview with the authors, August 2022.

[28] Government of Canada, 2010.

[29] Government of Canada, "Governor General's Special Warrants," webpage, updated October 19, 2015.

[30] Meagan Campbell, "How Canada Avoids U.S.-Style Government Shutdowns," *Maclean's*, January 23, 2018.

[31] Government of Canada, 2010.

FIGURE 3.3

Canada's Parliamentary Supply Periods and Expenditure Cycle

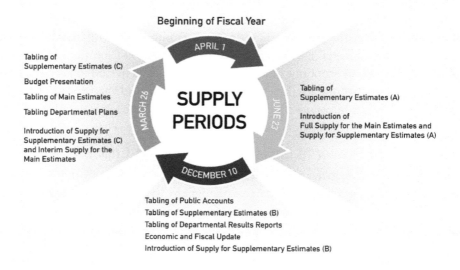

EXPENDITURE CYCLE

Beginning of Fiscal Year

APRIL 1

SUPPLY PERIODS

MARCH 26

JUNE 23

DECEMBER 10

Tabling of
Supplementary Estimates (C)

Budget Presentation

Tabling of Main Estimates

Tabling Departmental Plans

Introduction of Supply for
Supplementary Estimates (C)
and Interim Supply for the
Main Estimates

Tabling of
Supplementary Estimates (A)

Introduction of
Full Supply for the Main Estimates and
Supply for Supplementary Estimates (A)

Tabling of Public Accounts
Tabling of Supplementary Estimates (B)
Tabling of Departmental Results Reports
Economic and Fiscal Update
Introduction of Supply for Supplementary Estimates (B)

SOURCE: Reproduced from Government of Canada, 2010.
NOTE: Letters in parentheses identify the relevant Supplementary Estimates (A, B, or C), with A tabled in May, B in late
October or early November, and C in February. The appropriations act associated with each estimate becomes law
approximately one month after tabling.

Departmental Plans and Results Reports

All Canadian federal departments and agencies prepare annual departmental plans (DPs) and DRRs. These reports are "designed to reflect the government's citizen-focused agenda by identifying the benefits the department provides to Canadians and the real value derived for each taxpayer dollar spent."[32]

DPs outline individual expenditure plans for each department and agency over a three-year period, linking the organization's main strategic priorities and expected program results to the Main Estimates presented to parliament. Federal departments present their plans according to the government's Policy on Results,[33] which lays out a departmental results framework that sets forth each department's core responsibilities, expected results, and result indicators. DPs, which are tabled by the president of the Treasury Board soon after the tabling

---

[32] Government of Canada, Global Affairs Canada, "Planning and Performance," webpage, updated March 3, 2022.

[33] Government of Canada, "Policy on Results," webpage, updated July 1, 2016a.

of the Main Estimates, give parliament insight on planned spending and support the deliberation of the supply bills.[34] Once tabled, the plans are referred to parliamentary committees.[35]

DRRs are annual reports prepared by the agencies and departments. They are accounts of actual performance in the most recent fiscal year relative to the plans, priorities, and expected results that had been set out in the prior DPs.[36]

For example, in its FY 2021–2022 DRR, DND sought to achieve 93 results through its execution of six core responsibilities and 54 programs. Progress was measured using 153 indicators showing whether (1) an outcome had been met or not met or (2) the result was not available or was still to be achieved. Across those 153 indicators, DND met its targets 50 times (32.7 percent of indicators), did not meet its targets 52 times (34.0 percent of indicators), could not produce results 21 times (13.7 percent of indicators), and anticipated achieving results in 30 cases (19.6 percent of indicators).[37]

Overview of Canada's Defense Budgeting Process

In this section, we discuss DND's accounting practices, spending levels, decisionmakers and stakeholders, planning and programming, budgeting and execution, and oversight.

Accounting

Beyond adhering to Canada's federal budget process, DND uses both cash and accrual accounting.[38] According to the 2017 *Strong, Secure, Engaged* defense policy document, DND has operated two separate budgets as "a vestige of history."[39] DND funding, like that of all Canadian federal departments, is allocated on a cash basis through the parliamentary estimates process. But in 2005–2006, a separate part of DND's budget was created using accrual accounting to allow for long-term, predictable sources of funds for the acquisition, operation, and maintenance of major new equipment and to support force expansion. The result has

---

[34] Statistics Canada, "Departmental Plan," webpage, updated March 2, 2023.

[35] Government of Canada, "Departmental Results Reports," webpage, updated November 21, 2016b.

[36] Government of Canada, 2016b.

[37] Government of Canada, GC InfoBase, "Infographic for National Defence: Results," webpage, December 2, 2022.

[38] Under the accrual basis of accounting, "expenses for goods and services are recorded before any cash is paid out for them" (Chizoba Morah, "Accrual Accounting vs. Cash Basis Accounting: What's the Difference?" webpage, Investopedia, March 19, 2023). Thus, under this accounting practice, the government of Canada expenses the cost of an asset (e.g., a ship) starting when that asset enters service and spreads the cost of the asset over its useful life rather than noting it when the bills are paid. See, for example, Morah, 2023, and DND, 2017, pp. 96–100.

[39] DND, 2017, p. 44.

been that planning for the capital program is managed on an accrual basis, while operating costs are largely covered by the cash budget.[40]

DND Spending Levels

DND's expenditure plan for FY 2022–2023 had proposed approximately Canadian dollar (CAD) $25.9 billion, or roughly U.S. $19.4 billion. That figure included nearly CAD $24.3 billion (U.S. $18.2 billion) in voted expenditures—essentially, discretionary spending—and a little more than CAD $1.6 billion (U.S. $1.2 billion) in statutory spending. Total proposed spending rose by about CAD $1.6 billion (U.S. $1.2 billion) from the FY 2021–2022 proposal but fell by nearly CAD $1 billion (nearly U.S. $750 million) from FY 2020–2021 expenditures. Tables 3.2 and 3.3, later in this chapter, detail DND spending in the past three fiscal years and FY 2022–2023 proposed spending by purpose.

Canada expects to increase its defense spending by CAD $8 billion over the next five years to "bolster the capacity of the Canadian Armed Forces, support [CAF] members, and promote culture change." The expected increase includes CAD $7.4 billion to "increase defence capabilities, improve continental defence, and support commitments to [Canada's] allies."[41]

Decisionmakers and Stakeholders

DND is led by a single political appointee, a member of parliament from the party in power, who serves as the Minister of National Defence. The minister oversees the entirety of the defense portfolio. However, there may be up to two additional political appointees who carry titles associated with DND but who have little to no actual policymaking or decisionmaking authority within the department. The first, the Associate Minister of National Defence, is also a member of parliament who typically holds another cabinet position. There is no legal requirement that the associate ministerial role be filled; prime ministers since the 1960s have chosen to either fill or not fill the position as they see fit. Since 2015, under the current prime minister, the associate minister position has been held concurrently by the Minister of Veterans Affairs.

The second additional position is the parliamentary secretary to the Minister of National Defence. Parliamentary secretary positions allow the government to reward loyal backbench members of parliament, but they are not members of the cabinet, and they do not typically have policy portfolios within the department. Their primary roles are to answer questions in the House of Commons or to table reports on behalf of ministers when the ministers cannot be present.[42] Figure 3.4 reproduces the DND organizational chart.

---

[40] DND, 2017, p. 44.

[41] Parliament of Canada, House Standing Committee on National Defence, "Defence Spending: Budget 2022," webpage, updated April 27, 2022.

[42] Government of Canada, Privy Council Office, *Guide for Parliamentary Secretaries*, December 2015.

FIGURE 3.4
DND Organizational Chart

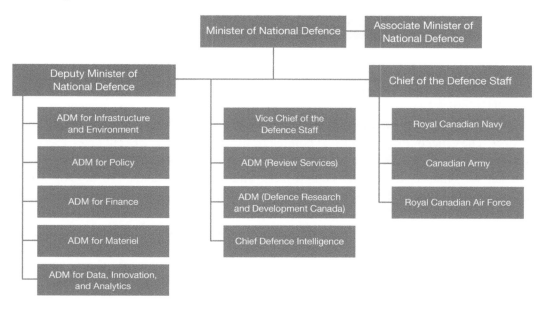

SOURCE: Adapted from DND, "DND/CAF Organizational Chart," webpage, February 23, 2022c.
NOTE: ADM = assistant deputy minister.

The Minister of National Defence has political responsibility for DND; the minister guides policy as set by the government and laid out in the governing party's election platform. The Deputy Minister of National Defence, in contrast, is the senior civil servant within the department and has ultimate responsibility for implementing the minister's policies. The deputy minister also has ultimate fiscal responsibility for DND and oversees several civilian assistant deputy ministers who are senior members of the civil service. These assistant deputy ministers oversee various portfolios or groups, including policy, materiel, and finance.

DND's Assistant Deputy Minister for Finance is the chief financial officer charged with leading the department's Finance Group, preparing the department's budget, and liaising with other federal agencies (specifically, the Treasury Board Secretariat and the Department of Finance). The Materiel Group oversees procurement, in-service support, military maintenance projects, and the implementation of acquisition projects.[43] Each of the military services has a comptroller who has a role in the programming, resourcing, and budgeting processes.[44]

CAF is led by the Chief of the Defence Staff, who reports to the Minister of National Defence and is the only four-star officer in the military. The services are not separate legal entities. (Until 2011, the services were referred to as *commands*; for example, today's Royal

---

[43] Canada subject-matter expert, interview with the authors, August 2022.

[44] Canada subject-matter expert, interview with the authors, December 2022.

Canadian Navy was known as the Maritime Command.) The services are led by three-star officers, while the commands (other than the services) are led by two-star officers. No civilian officials or political appointees lead the military services, and service comptrollers are also military officials. CAF is regarded as a single entity, so there is more jointness in resourcing and decisionmaking than in the United States. Still, each of the services has a role in the budget process and can advocate for its preferred procurement projects.

A military official serves as the Chief of Programme under the Vice Chief of the Defence Staff. This official leads an analytical process that ranks military procurement projects according to DND priorities, thereby signaling the likely apportionment of resources to the services. The Chief of Programme is appointed from among the services but takes a service-agnostic viewpoint in making decisions. This chief works with the DND's Assistant Deputy Minister for Finance, the service comptrollers, and other stakeholders to develop a key capabilities list for capital projects reflecting military priorities that were analyzed from a joint force perspective. This list is protected information that is not releasable to the public.[45] Should a military official seek to share the key capabilities list with the House Standing Committee on National Defence, the military official would most likely not find a receptive audience, given Canada's governing structure and a political climate that deemphasizes defense spending.[46] Furthermore, there is no earmarking or other legislative process that allows individual MPs to impose spending requirements on DND (or other Canadian federal agencies) that the agencies have not requested. As discussed previously, the executive has a significant amount of power in the budgeting process, as long as the executive's party commands a majority in parliament.

Planning and Programming

Planning and Policy

Canada's defense programs are based on several strategic planning documents: *Strong, Secure, Engaged* (2017); *Defence Plan, 2018–2023* and *Defence Investment Plan* (2018); updated *Defence Investment Plan* (2019); *Defence Capabilities Blueprint* (2020); and the *Department of National Defence and Canadian Armed Forces Departmental Plan*, the most recent of which was released in 2022. Whereas *Strong, Secure, Engaged* explains Canada's defense policy, the subsequent documents implement it. Together, these strategic documents provide the basis for defense budgeting decisions.

As noted, *Strong, Secure, Engaged* is Canada's defense policy. Released in June 2017, it identifies 281 previously approved projects and 52 new projects totaling CAD $108 billion (approximately U.S. $80.3 billion) over a 20-year period.[47] It also forecasts total defense spending (including acquisition, operating, and sustainment costs for new equipment) to

---

[45] Canada subject-matter expert, interview with the authors, August 2022.

[46] Canada subject-matter expert, interview with the authors, August 2022.

[47] DND, 2017, p. 102.

reach 1.4 percent of Canada's GDP by 2024–2025, which is still short of the 2-percent NATO target.[48]

The 2017 policy also emphasizes the government's dedication to a transparent model for defense budget planning based on the "rigorous, evidence-based analysis of Canada's defence needs and the resources required" to fulfill them.[49] It was the first time the policy was published in a formal document that provided a 20-year view of the defense budget.[50] (Previous policy statements had appeared as white papers and had projected defense spending out about two to three years or the remaining term of the ruling government.)[51] The 2017 policy further indicated that DND would use accrual accounting to manage funding for the acquisition of all capital assets over their expected lives and use cash accounting to manage equipment operating and sustainment costs.[52]

Following the publication of *Strong, Secure, Engaged*, DND released *Defence Plan 2018–2023*.[53] This 2018 plan divided the 20-year horizon into three time frames: Horizon 1 (1–5 years), Horizon 2 (5–10 years), and Horizon 3 (10–20 years). It outlined the strategic results that DND and CAF aimed to achieve in Horizon 1 by FY 2022–2023 and listed five defense priorities toward which DND and CAF leadership should direct resources to achieve those results.[54] Table 3.1 lists the planned strategic results and defense priorities from the plan. DND intends to update the plan "as required to reflect any changes or enhancements" following a review by the Vice Chief of the Defence Staff.[55]

Also in 2018, DND publicly issued its *Defence Investment Plan*. That plan's primary purpose was to inform policymakers, industry, defense experts, media, academics, and the public about the progress DND was making on its capital investments. The investment plan listed more than 200 capital defense projects, each costing more than CAD $5 million (approximately U.S. $3.7 million), and support contracts valued at more than CAD $20 million (approximately U.S. $14.9 million) that were expected to be awarded. The plan excluded completed projects and service contracts that were already underway. According to the 2018 investment plan, DND was striving to reduce its internal project development and approval

---

[48] DND, 2017, p. 6; J. Craig Stone, *Growing the Defence Budget: What Would Two Percent of GDP Look Like?* Canadian Global Affairs Institute, March 2017.

[49] DND, 2017, p. 43.

[50] Canada subject-matter expert, interview with the authors, October 2022; David Perry, "Canadian Defence Budgeting," in Thomas Juneau, Philippe Lagassé, and Srdjan Vucetic, eds., *Canadian Defence Policy in Theory and Practice*, Palgrave Macmillan, 2020, p. 65.

[51] Canada subject-matter expert, interview with the authors, October 2022.

[52] DND, 2017, p. 44.

[53] DND, *Defence Plan: 2018–2023*, 2018b.

[54] DND, 2018b.

[55] DND, 2018b, p. 1.

TABLE 3.1

DND and CAF Planned Strategic Results and Defense Priorities

DND and CAF Strategic Results	Defense Priorities
Canadians are protected against threats.	Achieve Canada's new vision for defense.
CAF is ready to conduct concurrent operations.	Foster well-supported, diverse, and resilient personnel and families.
DND and CAF have a diverse, resilient, and qualified workforce.	Grow and enhance capability and capacity.
Defense capabilities are designed to meet future threats.	Exploit defense innovation.
Defense procurement is streamlined and well-managed.	Modernize the business of defense.
Navy, Army, and Air Force installations are well-managed and enable military and defense activities.	
DND and CAF develop and maintain a business continuity plan.	
A defense security program reduces security risks in an evolving threat environment.	
Initiatives directed by the government of Canada are executed.	

SOURCE: Features information from DND, 2018b, pp. 6–7.

time by at least 50 percent for low-risk, low-complexity projects "through improved internal communication, increased delegation, and more efficient departmental approval processes."[56]

The 2018 investment plan was based on a technical plan that DND had submitted to the Treasury Board for review. This technical plan was condensed into the *Defence Capabilities Blueprint*, an online tool that "provides industry with access to information about defence investment opportunities."[57] Released alongside the 2018 investment plan, the blueprint detailed more than 200 projects across 13 broad land, sea, air, space, and cyber defence capability areas funded under *Strong, Secure, Engaged* and expected to be awarded in the coming years. The blueprint outlined funding ranges, project timelines, and links to project websites where applicable. The Treasury Board approved an updated investment plan and blueprint in 2022. The *Defence Capabilities Blueprint* is updated monthly to reflect recent project updates, contract awards, or other public announcements.[58]

The 2017 defense policy indicated that DND would refresh the investment plan annually and publish a new one every three years. Accordingly, DND updated the 2018 plan in 2019.

---

[56] DND, *Defence Investment Plan 2018: Ensuring the Canadian Armed Forces Is Well-Equipped and Well-Supported*, 2018a, p. 13.

[57] DND, 2018a, p. 8.

[58] DND, "Defence Capabilities Blueprint," webpage, updated January 9, 2020a.

At the time of this writing, the release of the 2022 update was on hold while DND assessed "the best timeframe to publish the next Defense Investment Plan to take into consideration the upcoming Defense Policy update."[59] The upcoming defense policy update will also incorporate NORAD modernization projects.

Finally, DND—like all Canadian federal agencies—releases its DP for the coming fiscal year and its DRR for the prior fiscal year, most recently in its *Department of National Defence and Canadian Armed Forces 2022–2023 Departmental Plan.*[60] The DP lays out how DND will meet its core responsibilities and details the department's strategic priorities, program activities, planned spending, and expected results over a three-year period.[61] The DP also provides metrics that are intended to measure the success or failure of efforts to meet the stated priorities. Despite the financial details and results provided in the DP and DRR, some Canadian scholars have found this information to be of little use because the documents "employ arbitrary categories that do not appear to connect to actual activities or spending accounts."[62]

DND publishes its DPs annually, regardless of whether a new government has issued a new defense policy. For example, the Liberal Party government that came to power in 2015 did not issue its defense policy until 2017, but DND still released its 2016 DP (likely to align more with the Liberal Party's priorities than with those of the prior Conservative Party's government).

Programming and Procurement

Guided by the strategic plans and policies discussed in the preceding section, DND initiates the process to acquire military equipment to meet capability deficiencies or emerging requirements identified by CAF. Canada's military equipment acquisition process follows the five phases illustrated in Figure 3.5.

In phase 1, the services identify a deficiency in CAF capabilities to meet current or future operational requirements.[63] This phase focuses on confirming the need—not the solution—and on verifying that the proposed requirements align with the issued policies and investment plans. The project team lists the capabilities that the acquisition should deliver, proposes a date by which the capabilities should be delivered, and specifies the earliest date on which the project would be ready to deploy. The project team must also show that no solution exists

---

[59] DND, "Defence Investment Plan 2018," webpage, updated December 9, 2022e.

[60] DND, *Department of National Defence and Canadian Armed Forces, 2022–2023: Departmental Plan*, 2022a.

[61] DND and CAF core responsibilities include operations, ready forces, defense team activities, future force design, capability procurement, sustainable bases, information technology systems, and infrastructure. See DND, 2022a; DND, 2017; and Shaowei Pu and Alex Smith, *The Parliamentary Financial Cycle*, Library of Parliament, Publication No. 2015-41-E, September 24, 2021.

[62] Perry, 2020, p. 66.

[63] DND, "Defence Purchases and Upgrades Process," webpage, updated September 10, 2018c.

FIGURE 3.5

Five Phases of Canada's Defense Acquisition Process

SOURCE: Adapted from DND, 2018c.

for the identified requirements. The Defence Capabilities Board (DCB) and the Independent Review Panel for Defence Acquisition (IRPDA) review the high-level requirements.[64]

As a point of comparison, DCB plays some similar roles to those of DoD's Joint Requirements Oversight Council. DCB oversees the identification and options analysis phases (phases 1 and 2) for all projects over CAD $5 million. Projects over CAD $100 million undergo additional, third-party review by the IRPDA. DCB reviews and vets the mandatory requirements of major projects and any rationale for selecting a proposed preferred option.[65] Specifically, DCB reviews a proposed project's strategic context document, which identifies the project's place in Canada's strategic framework and its capability requirements, including high-level mandatory requirements and viable options to be assessed. Broadly, DCB serves as a challenge function to ensure the alignment of future capabilities with strategies, the comprehensiveness of the options analysis, and the transparency and detail of the project costing figures. DCB also helps with the prioritization of capabilities out to five to 20 years. Unlike DoD's Joint Requirements Oversight Council, DCB reviews all DND programs across all services, not just those of special interest or with a joint role. Similar to its closest U.S. counterpart, DCB does not appear to be involved in creating capability requirements but rather in ensuring that proposed projects are aligned with existing requirements.

DCB is staffed by senior DND and CAF officials who are appointed by the Minister of National Defence, although we did not find a list of panel members. The IRPDA was created in 2015 and is made up of five independent, external appointees who advise DND senior leadership.

In phase 2, the project team prepares a statement of operational requirements and a business case analysis of options that would meet the requirements, along with a best or preferred option. The DND senior leadership uses the results of this analysis to identify the best option to meet the need while demonstrating value for money. DCB and the Program Management Board review the business case analysis results, challenge any rough-order-of-magnitude estimates, and provide senior internal management with support and advice. The IRPDA also

---

[64] DND, 2018c.

[65] IRPDA reviews and vets the requirements of major projects and the rationale for the selection of the preferred option based on business case analyses and the statement of operational requirements (DND, 2018a, pp. 13–15).

reviews and validates the preliminary operational requirements statement.[66] DCB endorses project progression into the next phase. The Program Management Board directly grants DND allocation authority for projects up to CAD $50 million (U.S. $37 million) and endorses projects over that amount.[67]

During these first two phases, the services or other DND components use their vote 1 operating funds to define new requirements and promote new projects.[68] In these phases, a service is investing in itself prior to a project's approval by the Minister of National Defence or the Treasury Board and receipt of funding through vote 5 capital expenditures and the Capital Investment Fund.[69]

In phase 3, DND begins to plan for a project that is affordable and achievable. The project team transitions from focusing on *what* to do to meet the needed capability to *how* to do it.[70] The team prepares a detailed description of the project, finalizes the statement of operational requirements, validates project costs, and prepares a project management plan. If the Program Management Board approves these plans, the project team submits its plans to the Minister of National Defence or the Treasury Board (depending on the level of spending required) to receive expenditure authority to proceed to the next phase.[71] If the Program Management Board does *not* grant approval, the project stalls in the definition phase. At this phase, approved projects have had their life-cycle costs accounted for and are transitioned from vote 1 funding to a vote 5 capital expenditure.

In phase 4, the project team implements the plans submitted in phase 3. The team receives permission to spend funds to build the capability. It then contracts for goods and services and is responsible for ensuring that the project remains "within scope, on time and within budget."[72] The team works with Public Services and Procurement Canada, a ministerial department that acquires defense goods and services on behalf of CAF and other government departments and has contracting authority for projects over CAD $5 million.[73] Armed with a solid business case analysis and an evaluated industry bid, the team conducts complex equipment tests with DND and CAF stakeholders and delivers military equipment according

---

[66] DND, 2018c.

[67] DND, *Defence Investment Plan 2018: Annual Update 2019—Ensuring the Canadian Armed Forces Is Well-Equipped and Well-Supported*, 2019, p. 12.

[68] Vote 1 has nothing to do with phase 1, and vote 5 has nothing to do with phase 5. Rather, the vote numbers roughly correspond to different colors of money and inform parliamentary appropriations decisions. For DND, vote 1 corresponds to spending on operating expenditures, and vote 5 corresponds to spending on capital expenditures. We discuss vote numbers in greater detail in the next section.

[69] Canada subject-matter expert, interview with the authors, August 2022.

[70] DND, 2018c.

[71] DND, 2018c.

[72] DND, 2018c.

[73] Government of Canada, "Defence and Marine Procurement," webpage, updated November 23, 2021b; Government of Canada, "Defence Procurement Strategy," webpage, updated November 3, 2021a.

to the schedule in the contract. CAF members train to use any equipment acquisitions. This training may take months or years, depending on the complexity of the new equipment.[74]

Phase 5 begins once the new capability and supporting organization are fully operational.[75] The project team sends a formal notification to DND senior leaders that the operational capability has been achieved and that any lessons learned have been recorded and disseminated. This phase typically lasts for about three months.

This five-phase process is the same for all military equipment acquisitions, regardless of type, scope, or complexity. Observers and practitioners of this acquisition process have noted that a "one-shoe-fits-all" approach is risk-averse and focuses more on process than results.[76] This process may not be agile or responsive enough in such areas as cybersecurity, given the fast pace of technology development and potential system vulnerabilities.[77]

Average time frames for each phase do not appear to be published, except for those that have been mentioned previously. Anticipated time frames for new and in-progress acquisition projects are included in the *Defense Capabilities Blueprint*, however. A selection of projects is included in Table 3.2. These projects cross capability areas, including air, emerging technology, real property, sea, space, and training and simulation. They also range in complexity and cost from CAD $50 million for mess and accommodation construction to CAD $56–60 billion for the surface combatant project. The options analysis and definitions phases are anticipated to occur within three to four years of project approval, but the implementation (i.e., construction and delivery) phase may last far longer.

Budgeting and Execution

When the Minister of Finance presents the annual national budget to the House of Commons, and when the president of the Treasury Board tables the Main Estimates of detailed spending, the estimates are broken down into one or more votes that correspond roughly to different colors of money. Each color of money is assigned an arbitrary, noncontiguous vote number. Common types of votes include operating votes, which are used to fund day-to-day operating costs; capital votes, which are used to procure assets with ongoing uses; and grants and contributions votes, which are used to transfer funds to other organizations or governments.[78]

---

[74] DND, 2018c.

[75] DND, 2018c.

[76] Canada subject-matter expert, interview with the authors, October 2022; Canada subject-matter expert, remarks at a RAND seminar, October 27, 2022.

[77] Canada subject-matter expert, remarks at a RAND seminar, October 27, 2022.

[78] Pu and Smith, 2021.

TABLE 3.2

Select Anticipated Acquisition Phase Time Frames

Phase	Arctic Over the Horizon Radar (NORAD modernization)	Advanced Improvised Explosive Device Detection and Defeat	Joint Deployable HQ and Signal Regiment Modernization	Construct Mess, Dining, and Accommodations Canadian Forces Base Bagotville	Canadian Surface Combatant	Defence Enhanced Surveillance from Space (NORAD modernization)
Anticipated start of phase 2: options analysis	2023–2024	Complete	Complete	Complete	Complete	In progress
Anticipated start of phase 3: definition	2025–2026	2023–2024	2023–2024	2022–2023	In progress	2024–2025
Anticipated start of phase 4: implementation	2027–2028	2025–2026	2025–2026	2024–2025	2023–2024	2028–2029
Anticipated initial delivery	2028–2029	2026–2027	2027–2028	2026–2027	Early 2030s	Beyond 2035
Anticipated final delivery	2031–2032	2029–2030	2031–2032	2027–2028	2040s	Beyond 2035

SOURCE: Features information from DND, "Defence Capabilities Blueprint," webpage, updated December 1, 2022d.

NOTE: HQ = headquarters.

For instance, DND's FY 2022–2023 budget estimate contained four votes:

- vote 1, by far the largest category, for operating expenditures
- vote 5 for capital expenditures, including major capability procurement programs and infrastructure projects
- vote 10 for grants and contributions, including payments to NATO and funding for partner-nation military programs
- vote 15 for long-term disability and life insurance plans for CAF members.

Table 3.3 shows a detailed breakdown of DND's FY 2021–2022 and 2022–2023 Main Estimates.

Included with the Main Estimates is an alternative breakdown of spending by purpose, including such items as ready forces, procurement of capabilities, Defence Team, and future force design.[79] One scholar noted that the purposes provided in the breakdown (as seen in Table 3.4) can vary annually, making it difficult to track spending over time.[80]

---

[79] Government of Canada, 2022b, Table 181.

[80] Canada subject-matter expert, interview with the authors, December 2022.

TABLE 3.3

Organizational Estimates of DND Spending, FYs 2021–2022 and 2022–2023

Spending Category	FY 2021–2022			FY 2022–2023 Main Estimates
	Expenditures	Main Estimates	Estimates to Date	
Budgetary vote				
Vote 1. Operating expenditures	16.84	16.45	17.46	17.57
Vote 5. Capital expenditures	4.95	5.70	5.80	5.96
Vote 10. Grants and contributions	0.26	0.25	0.31	0.31
Vote 15. Payments for long-term disability and life insurance plans for CAF members	0.42	0.42	0.53	0.45
Total appropriations voted	22.47	22.82	24.10	24.29
Total statutory funding	4.35	1.48	1.64	1.66
Total funding	26.83	24.30	25.74	25.95

SOURCE: Features information from Government of Canada, 2022b, Table 180.

NOTES: All amounts shown are in CAD $ billions. Estimates to date include Supplements A, B, and C. Voted categories are funded through appropriations bills. *Statutory funding* refers to expenditures authorized by parliament through other legislation.

Voted appropriations can span a portfolio of programs or apply to specific programs. Examples of DND portfolios of programs are the Armored Combat Support Vehicle Fleet, Compensation and Benefits, Remotely Piloted Aircraft Systems, and the Canadian Surface Combatant Project. An example of a specific program that falls into this category is the CF-18 Hornet Extension Project.[81] However, because large votes include a variety of items, it is often impossible for someone outside DND to know how much an organization spends on a single program.[82] Organizations can also transfer funds within a vote from one program to another without parliament's approval based on trade-off analysis rather than set dollar thresholds.[83] Organizations do need parliament's approval to transfer funds *between* votes (e.g., from vote 1 to vote 5). Canadian federal agencies can also carry forward a portion of their unspent funds from a previous year, typically up to 5 percent of operating expenditures and 20 percent of capital expenditures.[84]

---

[81] DND, "Supplementary Estimates (A)—National Defence," webpage, updated April 7, 2020b.

[82] Raphaëlle Deraspe, *Funding New Government Initiatives: From Announcement to Money Allocation*, Library of Parliament, Publication No. 2021-32-E, October 7, 2021; Canada subject-matter expert, interview with the authors, November 2022.

[83] Pu and Smith, 2021.

[84] Pu and Smith, 2021.

TABLE 3.4

Organizational Estimates of DND Spending, FY 2022–2023

Budgetary Purpose	Operating	Capital	Transfer Payments	Revenues and Other Reductions	Total
Ready forces	10.02	0.53	0.003	(0.10431)	10.45
Procurement of capabilities	0.58	4.21	—	(0.00001)	4.79
Sustainable bases, information technology systems, and infrastructure	3.43	0.84	0.037	(0.17442)	4.13
Defence Team[a]	3.72	0.03	0.005	(0.01605)	3.74
Future force design	0.51	0.28	0.032	(0.00037)	0.82
Operations	0.51	0.05	0.239	(0.00002)	0.79
Internal services[b]	1.21	0.02		(0.01287)	1.22
Total	19.99	5.60	0.315	(0.30803)	25.95

SOURCE: Features information from Government of Canada, 2022b, Table 181.

NOTE: All amounts shown are in CAD $ billions.

[a] Defence Team is the spending category that includes initiatives to support Canada's military and civilian defense personnel and their families, including health services, human resources planning, family resilience resources, and programs to strengthen diversity, equity, and inclusion.

[b] The category of internal services includes expenditures in ten service categories: acquisition management, communication, financial management, human resources management, information management, information technology, legal, material management, management and oversight, and real property management.

The DP and DRR provide public assessments of DND programs and goals, strategic outcomes, and planned and expected results. DND has created an internal system to track funding for individual programs or offices, but that level of detail is not available in the annual budget or in the estimates for *initial* parliamentary or public review. Two parliamentary oversight entities—the Auditor General and the Parliamentary Budget Officer—receive that level of detail upon request to conduct audits or spending reviews to inform *subsequent* parliamentary decisions or departmental management.[85] Likewise, DND's internal Review Services division may have access to that level of detail for internal auditing purposes.

Oversight

Some commentators have stated that legislative oversight of Canada's defense and military affairs is limited, even suggesting that the term *scrutiny* is more accurate than *oversight* to describe parliament's role in reviewing defense spending.[86] Pointing to a structural reason for this limitation, Philippe Lagassé, a professor at the School of International Affairs at Carleton

---

[85] Canada subject-matter expert, interview with the authors, November 2022.

[86] Canada subject-matter expert, interview with the authors, December 2022.

University, argues that "the Canadian Parliament is particularly weak" compared with the British and Australian parliaments. He notes that British parliamentarians are willing to hold governments to account, in part owing to more backbench independence and a more "forceful" committee structure.[87] The Australian Senate, according to Lagassé, performs more legislative and parliamentary scrutiny of the armed forces than does the Canadian Senate, which "is no longer as seized of national security and defence as it was in previous decades."[88] Both British and Australian legislative oversight pales in comparison to the power and influence of the U.S. Congress, but according to Lagassé, Canada is particularly hobbled by strong party discipline, excessive partisanship, the power of party leaders, and high member turnover.[89]

During its periodic review of estimates, the Canadian Parliament may spend around ten hours *annually* scrutinizing spending, according to a briefing developed by Canadian Global Affairs Institute president David Perry.[90] He reviewed the six meetings held by two House of Commons committees—the Standing Committee on National Defence and the Standing Committee on National Finance—to examine DND's CAD $20.5 billion (U.S. $15.2 billion) budget for FY 2017–2018. The Committee on National Defence held three meetings, lasting up to two hours each, on the Main Estimates and Supplementary Estimates B and C. (That year, roughly 90 percent of DND's budget was requested during the Main Estimates period.) The Committee on National Finance held three meetings as well, two of which reviewed spending by DND and by other departments. None of that committee's meetings lasted more than 90 minutes.[91]

In 2022, the Committee on National Defence held two meetings to review DND spending: one reviewing Supplementary Estimates C for FY 2021–2022, which occurred in late March and lasted for nearly two hours, and another reviewing FY 2022–2023 Main Estimates, which occurred in June and also lasted for two hours. As a point of comparison, the Committee on National Defence held meetings on Arctic security and on recruiting and retention that cumulatively lasted about 20 hours and 7.5 hours, respectively.[92]

Parliamentary oversight—or scrutiny—in Canada is aided by analyses from the Auditor General, the Parliamentary Budget Officer, and, at times, the Library of Parliament. The former two roles are accountable directly to parliament rather than to the executive or a minister. The Auditor General is appointed by the governor in council (who is appointed by the governor general) following consultation with the leader of every recognized party

---

[87] Philippe Lagassé, "Improving Parliamentary Scrutiny of Defence," *Canadian Military Journal*, Vol. 22, No. 3, Summer 2022, p. 20.

[88] Lagassé, 2022, p. 20.

[89] Lagassé, 2022, p. 21.

[90] David Perry, "DND Spending: A View from the Outside," briefing slides, Canadian Global Affairs Institute, Executive Leaders Program, September 12, 2019.

[91] Perry, 2019.

[92] Parliament of Canada, House Standing Committee on National Defence, "Meetings, 44th Parliament, 1st Session," webpage, undated.

in the Senate and House of Commons and the passage of a resolution in both chambers. The Auditor General holds office for a ten-year term, issues an annual report to the House of Commons, produces other reports or audits on topics of the Auditor General's choosing during the year, and appears regularly before parliamentary committees.[93] The Parliamentary Budget Officer is appointed in a similar manner, holds office for a seven-year term, and provides estimates on matters relating to Canada's finances or economy either independently or at the request of a parliamentary committee. The Parliamentary Budget Officer issues an annual report to both chambers of parliament in addition to reports requested by committees or parliamentarians, all of which are meant to raise the quality of debate and promote budget transparency. At the beginning of each fiscal year, the Parliamentary Budget Officer also submits an annual work plan with a list of matters that the office intends to bring to the attention of parliament.[94]

Reports by the Auditor General and Parliamentary Budget Officer tend to cover broad topics, such as a review of the national shipbuilding strategy, CAF's supply chain and delivery times, Canada's defense expenditures and NATO's 2-percent target, and the life-cycle costs of surface combatants.[95] Notably, the Auditor General has reported on DND's difficulties in recording the quantities and values of its inventory for 16 years.[96] In 2020, an internal audit (likely conducted by DND's Review Services division) criticized DND management for assigning fewer than three people to monitor spending associated with the Liberal government's 2017 *Strong, Secure, Engaged* defense policy. The audit report, dated November 2019, came to light after DND revealed that more than 100 of the roughly 300 capital projects outlined in its policy were delayed. In response, DND officials said that some of the issues raised by the internal audit were being addressed but did not mention whether any additional staff were assigned to monitor spending.[97]

---

[93] Andre Barnes, *Appointment of Officers of Parliament*, Library of Parliament, Publication No. 2009-21-E, August 19, 2021, p. 2.

[94] Barnes, 2021, pp. 9–10.

[95] See Office of the Auditor General of Canada, *Report 2: National Shipbuilding Strategy: Independent Auditor's Report*, 2021; Office of the Auditor General of Canada, *Report 3: Supplying the Canadian Armed Forces—National Defence: Independent Auditor's Report*, Spring 2020; Christopher E. Penney, *Canada's Military Expenditure and the NATO 2% Spending Target*, Office of the Parliamentary Budget Officer, June 9, 2022; and Carleigh Busby, Albert Kho, and Christopher E. Penney, *The Life Cycle Cost of the Canadian Surface Combatants: A Fiscal Analysis*, Office of the Parliamentary Budget Officer, October 27, 2022.

[96] Office of the Auditor General of Canada, "The Auditor General's Observations on the Government of Canada's 2018–2019 Consolidated Financial Statements," webpage, undated.

[97] Lee Berthiaume, "Auditors Call Out National Defence for Poor Oversight on Spending," Global News, June 14, 2020.

Analysis of Canada's Defense Budgeting Process

Strengths

We divide the strengths of Canada's defense budgeting process between policy choices that help DND manage funds more efficiently and its budgeting mechanisms more generally.

DND's first policy strength is that Canada's government is never at risk of a shutdown because of funding lapses. Parliament can enact interim estimates that authorize spending at proposed levels until the Main Estimates pass through the normal legislative process, or the executive can take other extraordinary measures via the governor general to continue funding ongoing government functions.

The second policy strength is that DND's capabilities investment process allows the continual weighing of the marginal value of future capabilities to ensure that they remain credible and valuable.

The third policy strength is that Canada recognizes its relative military size on the world stage and its role as an "alliance dependent country."[98] To that end, Canada's defense budgeting policies emphasize cooperation with allies, specifically the United States and other NATO member countries, because DND "accept[s] the fact that [it is] not going to have everything."[99] Furthermore, Canada lacks the population and military personnel to sustain large overseas military deployments, and, thus, its 2017 policy limits the size and duration of planned contributions. Nonetheless, CAF participates throughout the year—again, largely with allies—in operations and joint military exercises, including assurance missions, stability operations, and United Nations missions.[100] Canada's recognition of the size and capacity of its military results in policies that focus on projects that can remain credible and valuable in the future.

The fourth policy strength is DND's internal financial transparency, which has been improving with better projections, better internal data management, and better analytics. Because Canada has a relatively small taxpayer base that funds its government programs, better data help DND justify increases in spending, help parliament better understand why those increases are necessary, and reassure the public that funds are spent as intended.[101] However, some observers of DND financial processes argue that financial transparency outside DND (i.e., to the public) has not improved enough; they want to see more financial transparency outside DND and the whole of government in general.[102] DND spending is still difficult for the public to track because the Main Estimates documents provided to parliament do not detail spending on similar categories over time. DND also does not publish spending

---

[98] Canada subject-matter expert, interview with the authors, October 2022.

[99] Canada subject-matter expert, interview with the authors, October 2022.

[100] For a list of operations and exercises, see DND, "Current Operations and Joint Military Exercises List," webpage, updated January 13, 2022b.

[101] Canada subject-matter expert, interview with the authors, October 2022.

[102] Canada subject-matter expert, interview with the authors, December 2022.

requests by program, as DoD does when it requests congressional appropriations. The Canadian public relies on the Auditor General or Parliamentary Budget Officer to critique DND spending and financial management and to suggest corrections.

The fifth and final policy strength is that DND manages its military acquisition projects through a service-agnostic process and ranks those projects according to DND capability priorities. This process ensures that service-centric views do not dominate procurement planning and encourages more collaboration, discussion, and consensus toward achieving Canada's defense plans and strategy. This process results in a strategy that develops, generates, and employs the force and its military power, which provides a push needed to move defense programs forward.

Furthermore, DND's budgeting process generates a key capabilities list that by law cannot be publicized, because the list represents plans that are close hold for DND. DND does not release this list—or any lists of unfunded priorities from the military services (e.g., Army, Air Force, Navy) that may disclose potential defense vulnerabilities—to parliament without express government departmental approval. DND considers this practice a strength because the absence of legislative interference at this level helps DND keep control over its strategy and budget while ensuring that projects align with needed capabilities.[103] Consequently, parliament is not able to second-guess military decisions, and individual parliamentarians cannot advocate for preferred, pork-barrel programs.

Overall, the Canadian Parliament has little political appetite to increase spending above what the executive proposes, which also limits Canada's ability to meet NATO's goal of spending 2 percent of GDP on defense.

In terms of budget mechanisms that are not tied to specific policy choices, DND's notional budget is guaranteed to continue year on year, allowing for better decisionmaking in out-years. Regular supplementary parliamentary spending periods can help close unforeseen funding gaps for emerging requirements and can help manage risk. DND officials believe that planning capital investments on an accrual basis while managing year-on-year funding on a cash basis allows for a more flexible funding model.[104] DND's Capital Investment Fund ensures that approved projects will be paid for years or even decades to come, regardless of a change in government.

DND also appears to have more flexibility than DoD in how it uses the funding it receives. DND does not require parliamentary approval—nor must it inform parliament—to transfer funds within a vote from one program to another. DND can carry forward to the next fiscal year up to 5 percent of total operating expenditures, which it can use to adjust misalignments in spending. Sometimes, of course, DND must return expired funding to the Ministry of Finance at the close of the fiscal year.[105]

---

[103] Canada subject-matter expert, interview with the authors, October 2022.

[104] Canada subject-matter expert, interview with the authors, October 2022.

[105] Dave Perry, *A Primer on Recent Canadian Defence Budgeting Trends and Implications*, School of Public Policy Research Papers, University of Calgary, Vol. 8, No. 15, April 2015.

Challenges

We divide the challenges in Canada's defense budgeting process into two categories as well: those external to the process but that influence it and those internal to the process.

We found several challenges external to the defense budgeting process. DND has little personnel capacity to absorb new funding or to improve process timelines. (An entity like DoD's Office of Cost Assessment and Program Evaluation [CAPE] might be helpful, but DND's entire financial staff is roughly the size of CAPE's staff.) Major procurement can take seven to ten years, which is not unique, but still hampers the deployment of new, necessary capabilities. The defense industry in Canada does not typically work on the government's fiscal year schedule, and fitting spending to production schedules can be complicated. Major contracting and FMS for DND (and other federal agencies) are handled by Public Services and Procurement Canada, which can cause delays. Barring a major shift in defense policy, DND is rarely able to cancel major investment plan programs, locking it into projects that may no longer be technologically cutting-edge or that have increased in cost. Finally, upgrades to existing equipment require the initiation of new procurement programs, although changes are underway to reform this process.

We also found several challenges internal to the defense budgeting process. The Treasury Board follows a one-size-fits-all approach to new projects, whether they are aircraft, information technology systems, or new infrastructure. Newly proposed capabilities can languish for two or three years in the definition phase (the third phase of the defense acquisition process), which involves confirming that funding is available and is being spent appropriately. Spending and other authorities are held at a high level and are typically not delegated; although, in recent years, the Minister of National Defence has begun delegating some capital investment authorities to the civil service Deputy Minister of National Defence. Accrual accounting for life-cycle planning can present challenges when taking into account inflation or exchange rates for FMS; these challenges can require revisions to cost estimates to improve affordability. Another internal challenge, though one not unique to Canada, is the bureaucratic inefficiency that may arise as the department aims to spend appropriated funds toward the end of the fiscal year, despite DND's ability to carry over some of its funds to the following year. Finally, the use of both cash and accrual accounting can cause some communication problems and public confusion when new spending is announced.

Applicability

Although Canada's defense budgeting process supports a parliamentary democracy that emphasizes alliances, diplomacy, and limited overseas military intervention, the Canadian government has made policy choices that could enhance efficiencies in other countries. For example, a unified CAF emphasizes jointness, rather than competition, between the services. DND's procurement process ranks capital projects across the services according to the needs of the entire DND. Individual services can develop and manage their own procurement projects up to a point and can participate in the prioritization process that determines when programs are funded.

As a result of Canada's political culture and policy decisions, there is an unfunded priorities list that is not released regularly to parliament. A list of funded and unfunded capital programs is developed internally, but it is not released publicly. This approach prevents legislative intervention in DND's budgeting process, allowing DND to (ideally) make and defend spending decisions according to its mission rather than politics.

Lessons from Canada's Defense Budgeting Process

Lesson 1: DND Political Leadership and CAF's Structure Promote Service-Agnostic Decisionmaking

Because Canada's military is relatively small compared with those of some of its allies (e.g., the United States), DND is led by a single political appointee (with assistance from civil public servants and the uniformed CAF). This leadership helps DND speak to the public with one voice and prevents different services from seeking a legislative champion for a preferred policy or spending authority when it is not internally approved. Furthermore, CAF is a single, unified legal entity led by a single four-star service member and acquires defense capabilities as one entity (i.e., jointly) as a rule and not as an exception. Although DoD may not be able to replicate this structure easily, a unified armed force may help improve jointness in planning and operations. A unified force might aid in other processes, including the procurement of capital projects.

Lesson 2: DND Sees Strengths in DoD's PPBE System

Our interviewees noted that the value in understanding DoD's PPBE System lies in understanding how Canada's biggest ally does business.[106] They mentioned two strengths in DoD's PPBE System from which U.S. allies and partners can learn. The first is the role of Congress. Congressional decisions allow for the release of funds to DoD programs and do not require any additional approvals from other government agencies outside DoD. However, in Canada, in addition to internal approvals, DND must seek Treasury Board, cabinet, and parliamentary approval to spend funds on new programs. Second, U.S. military services manage strategic deterrence. If DoD wants to preserve strategic deterrence at all costs, the military services will be required to generate the people, skills, and competencies to achieve this goal. Other countries may not be as equipped to prioritize their resources for that mission.

---

[106] Canada subject-matter expert, interview with the authors, October 2022.

Lesson 3: Certain Acquisition and Spending Problems Are Not Unique to DoD

DND faces long development times for acquisition programs. Some programs can spend up to three years in the definition phase, and it can take seven to ten years for major procurement programs to make their way through the acquisition process. These lengths of time can hamper the deployment of new, necessary capabilities, which can be obsolete once they reach service members. Although DND initiatives in recent years have made the process more efficient, a different set of reforms would be needed to acquire capabilities any faster. DND has also faced spending challenges in recent years. Although it has the authority to carry over up to 5 percent of its annual budget, this has not prevented DND from returning money above that limit that it has not been able to spend. DND's relatively small staff and budget, compared with those of DoD, limit its ability to absorb large influxes of money, which could partially explain why Canada's defense spending has not reached 2 percent of its GDP.

Lesson 4: Carryover Authorities Help Temper the Use-It-or-Lose-It Mentality

DND can carry over up to 5 percent of its annual operating expenditures to the next fiscal year. This flexibility can help reduce unnecessary end-of-year spending before authorizations expire. Still, this authority has not prevented DND from having lapsed funds or regularly spending less than its budget. In March 2022, the Parliamentary Budget Officer found that DND had underspent on capital investments since the release of *Strong, Secure, Engaged* in 2017.[107] Such lapses do not necessarily result in decreased spending plans for DND programs; in fact, DND plans to increase its capital expenditures starting in FY 2025–2026. Changes in project scope, delays in project schedules, and changes to cost estimates may, however, result in a need to adjust capital spending.

Lesson 5: Canada Has No Unfunded Priorities List and Little Legislative Interference

Canada's political structure does not allow parliament to drastically change funding for departments, including DND, beyond what has been requested. Canada's political culture means that there is typically not much appetite for large increases in DND's spending in any given year. Although DND maintains a list of capabilities it intends to procure, that list is not requested by—or even allowed to be passed to—parliament. This prohibition limits parliament's ability to second-guess military decisions just as it prohibits individual parliamentarians from advocating for preferred, pork-barrel programs.

---

[107] Christopher E. Penney and Albert Kho, *Planned Capital Spending Under Strong, Secure, Engaged—Canada's Defence Policy: 2022 Update*, Office of the Parliamentary Budget Officer, March 11, 2022.

Table 3.5 summarizes the lessons from Canada's defense budgeting process that may be relevant for DoD.

TABLE 3.5

Lessons from Canada's Defense Budgeting Process

Theme	Lesson Learned	Description
Decisionmakers and stakeholders	Lesson 1: DND political leadership and CAF's structure promote service-agnostic decisionmaking.	DND's unified CAF promotes jointness in planning and aids in the procurement of capabilities that align with overall DND priorities without favoring one service over another.
	Lesson 2: DND sees strengths in DoD's PPBE System.	Noted strengths of DoD's PPBE System include legislative approval and U.S. military services' ability to manage strategic deterrence.
Planning and programming	Lesson 3: Certain acquisition problems are not unique to DoD.	Long development and procurement timelines plague Canada's DND, despite its comparatively smaller size and budget.
Budgeting and execution	Lesson 4: Carryover authorities temper the use-it-or-lose-it mentality.	DND can carry forward 5 percent of its annual budget to the next fiscal year, reducing year-end pressure to spend funds.
Oversight	Lesson 5: There is no unfunded priorities list and little legislative interference.	Canada's political culture does not incentivize parliamentarians to interfere in the expenditure plans of departments or to increase DND spending above what is requested.

CHAPTER 4

United Kingdom

James Black, Nicolas Jouan, and Benjamin J. Sacks

The UK is a constitutional monarchy with a bicameral parliamentary system. The stability of the bicameral system relies on the fact that the chief of the executive branch (the prime minister, formally the First Lord of the Treasury) is a member of parliament from whichever party is able to command the confidence of a majority of the elected members of the lower chamber, the House of Commons.[1] Members of the upper chamber, the House of Lords, are not elected but appointed. The government of the day may or may not hold a majority in the House of Lords, whose function is largely to offer advice and scrutiny. By centuries-old convention, the upper chamber defers to the lower chamber on financial matters, limiting the upper chamber's ability to amend or block spending bills.

This interweaving of the executive and legislature, along with the use of a "first-past-the-post" (or plurality) voting system to elect MPs, is intended to empower the prime minister and the prime minister's chosen cabinet to rule with a strong mandate. Because the UK's government necessarily emerges from the parliament's majority, there is less inherent antagonism between the branches of government than in the United States. The resulting empowerment of the prime minister can enable more-streamlined executive and legislative action, but it also limits the formal checks and balances that characterize the U.S. system.

Within this structure, parliament must approve the resources that the MoD requests to perform its government-mandated missions. Without this approval, there are consequences for the prime minister: de facto opposition from the prime minister's own majority in the House of Commons, triggering a no-confidence vote and the likely collapse of the current government. The alignment of resource allocation with the MoD's mission is therefore a structural feature of the UK parliamentary system, at least as far as the government properly estimates the resources needed to satisfy its defense needs. In this chapter, we provide an overview of the MoD's approach to PPBE and insights that should be relevant to DoD.

---

[1] This can be via a party winning an outright majority of seats (as is typically the case), by a party entering a formal coalition with one or more other parties (as the Conservatives and Liberal Democrats did in 2010–2015), or through a looser arrangement known as *confidence and supply*, whereby a minority party rules alone but with other parties agreeing to back it on votes of confidence or supply, even if they do not enter into a formal coalition government.

PPBE in the Context of the MoD

The MoD's approach to PPBE begins with its mission as outlined in the Defence Command Paper. This white paper aligns the MoD's priorities with the broader Integrated Review of Security, Defence, Development and Foreign Policy, which is published by the prime minister's Cabinet Office.[2] The most recent iteration of the white paper, *Defence in a Competitive Age,* was published in March 2021, immediately after the Cabinet Office's latest integrated review focusing on the UK's overall global competitiveness.[3]

The white paper states that the seven primary goals of the MoD and the British Armed Forces are to defend the UK and its Overseas Territories, sustain the country's nuclear deterrence capacity, project the UK's global influence, execute its NATO responsibilities,[4] promote national prosperity, contribute to peacekeeping, and support the defense and intelligence-gathering capabilities of the UK's allies and partners.[5] HM Treasury aligns fiscal resources to support these missions through comprehensive spending reviews. Conducted every three to five years, these reviews provide an overall financial plan for the various parts of the central government, including the MoD.[6] Comprehensive spending reviews are then translated into corresponding departmental plans, and the departments' associated annual budgets are submitted to parliament.[7]

The structural alignment between the MoD's missions and the resources allocated to them does not mean that the estimated or allocated defense resources are always realistic. Indeed, since World War II and the decline of the British Empire, the UK has struggled with inherent tensions between its global commitments and ambitions on one hand and its resource limitations as a medium power on the other. Like the United States, the UK is a permanent member of the United Nations Security Council, a nuclear power, and a key player in

---

[2] The notion of a cross-governmental integrated review process was introduced in the wake of the UK's decision to leave the European Union, which underscored the need for a more integrated approach across all levers of policy and to define the new role of global Britain. Previously, the UK had occasionally undertaken narrower reviews, including the Strategic Defence Review of 1998, Strategic Defence and Security Reviews of 2010 and 2015, and the National Security Capability Review of 2018. Policymakers had planned to conduct integrated reviews every five years (with the next taking place around 2025), but Russia's invasion of Ukraine in February 2022 prompted a more limited refresh of the assumptions underpinning the 2021 review, which was expected to be completed in early 2023.

[3] MoD, *Defence in a Competitive Age*, March 2021a; UK Cabinet Office, *Global Britain in a Competitive Age: The Integrated Review of Security, Defence, Development and Foreign Policy*, March 2021.

[4] UK House of Commons, Defence Select Committee, "Memorandum for the Ministry of Defence: Supplementary Estimate 2021–22," March 2, 2021, p. 1.

[5] MoD, "About Us," webpage, undated-a.

[6] HM Treasury, "Autumn Budget and Spending Review 2021 Representations," webpage, September 7, 2021.

[7] MoD, *Financial Management and Charging Policy Manual, Part 2: Guidance*, version 7.0, Joint Service Publication 462, March 2019b, withdrawn November 27, 2020, p. 16; UK subject-matter experts, interviews with the authors, November 2022.

NATO. It also aspires to maintain the high-end, full-spectrum military capabilities needed for expeditionary operations around the globe, not least to protect its Overseas Territories.[8] Unlike the United States, however, the UK lacks the resources to train and equip large-scale forces, so it must focus on quality over quantity and look to interoperability and a network of alliances to provide additional force strength and global reach.[9]

Managing this tension between ambitions and limitations implies an interdependence on allied and partner resource decisionmaking and development timelines, at least outside select areas deemed essential to sovereignty. This interdependence is reflected in the emphasis in the MoD's 2021 *Defence and Security Industrial Strategy* on an *own-collaborate-access* approach to MoD investments and capability management.[10] The MoD aims to *own* critical sovereign capability through UK-only programs when strictly necessary; *collaborate* on joint programs with allies and partners where possible to spread costs and risk while achieving economies of scale for production, exports, and life-cycle support; and *access* cheaper, off-the-shelf products and services from the wider market wherever prudent.[11]

These financial tensions have been exacerbated by acute near-term pressures, both on the MoD's budget and on the agility and flexibility of its PPBE processes. Even before Russia's invasion of Ukraine in February 2022, the MoD and military were undergoing a period of sweeping modernization and transformation. The objective was to position the UK for (1) a more robust response to sub-threshold gray zone threats (e.g., from Russia, China, or Iran) and (2) a return to potential warfighting as part of NATO. The escalating conflict in Ukraine added new urgency, as well as new distractions and demands.

The UK's Defense Spending Ambitions

The UK government is committed to maintaining defense spending above 2 percent of GDP, in line with NATO targets. In its 2019 manifesto, the ruling Conservative Party also committed to increasing the MoD's budget by at least 0.5 percent above inflation every year of the

---

[8]   The 14 British Overseas Territories are spread across the globe and are Anguilla; Bermuda; the British Antarctic Territory; the British Indian Ocean Territory (home to the U.S. base at Diego Garcia); the British Virgin Islands; the Cayman Islands; the Falkland Islands; Gibraltar; Montserrat; the Pitcairn Islands; Saint Helena, Ascension, and Tristan da Cunha; South Georgia and the South Sandwich Islands; the Sovereign Base Areas of Akrotiri and Dhekelia in Cyprus; and the Turks and Caicos Islands. The UK also has bases in other countries, such as Canada, Belize, Kenya, Saudi Arabia, Oman, Bahrain, Nepal, Brunei, and Singapore.

[9]   MoD, 2021a.

[10]   MoD, 2021b.

[11]   RAND Europe was asked to develop a decision support tool for the MoD to help it navigate its choices using the own-collaborate-access model. See Lucia Retter, James Black, and Theodora Ogden, *Realising the Ambitions of the UK's Defence Space Strategy: Factors Shaping Implementation to 2030*, RAND Corporation, RR-A1186-1, 2022.

current parliament to insulate the ministry from any inflationary shocks.[12] In 2020–2021, the UK's defense expenditures nominally amounted to £42.4 billion ($48.07 billion), or just above 2 percent of GDP. Of this, roughly £0.5 billion ($0.57 billion) was spent on military operations, primarily in Afghanistan, and on overseas training.[13] This budget makes the MoD one of the largest UK government departments by expenditure (behind health, welfare and pensions, and education). It also makes the UK the second-biggest spender on defense among NATO member countries (behind the United States).[14]

Recently, however, the twin shocks of the COVID-19 pandemic and the war in Ukraine—and the resultant economic disruptions—have strained these commitments to defense spending. In her brief tenure as prime minister in September and October 2022, Liz Truss pledged to raise the MoD's budget to 3 percent of GDP in response to growing threats, a move that would have cost the country an extra £157 billion by 2030.[15] However, with the onset of an economic recession and the added financial instability triggered by the government's fiscal policies, this ambition for 3 percent was watered down under Truss's successor, Rishi Sunak.[16] The government has also been forced to revisit its prior commitment to keep increasing defense spending above the rate of inflation, which spiked to more than 10 percent in 2022. Thus, even a growing defense budget could be subject to cuts in real, if not nominal, terms.[17] Inflationary pressures have been exacerbated by a strengthening U.S. dollar, to which the MoD is especially sensitive because of its large number of major FMS contracts with the United States, as well as some fixed-price fuel-swap contracts that are denominated in dollars.[18]

Although it had not anticipated another full integrated review until 2025, the UK government was already conducting a "refresh" of the 2021 review in 2022,[19] in light of the "strategic shock" of Russia's invasion of Ukraine and heightened tensions with China. Published in March 2023, the refresh announced the injection of £5 billion of additional funding over two years to help the MoD address the impact of the war in Ukraine and high levels of inflation. It

---

[12] Conservative Party, *Get Brexit Done, Unleash Britain's Potential: Conservative and Unionist Party Manifesto 2019*, 2019.

[13] Esme Kirk-Wade, *UK Defence Expenditure*, House of Commons Library, April 6, 2022.

[14] MoD, "MOD Departmental Resources: 2021," webpage, February 24, 2022d.

[15] "Liz Truss Defence Spending to Cost £157bn, Says Report," BBC, September 2, 2022.

[16] Sebastian Payne and Sylvia Pfeifer, "Sunak Quiet on Defence Budget as He Signs Off on £4.2bn Frigate Contract," *Financial Times*, November 14, 2022.

[17] Dan Sabbagh, "Ben Wallace Steps Back from Liz Truss's 3% Defence Spending Target," *The Guardian*, November 10, 2022.

[18] The MoD maintains multiple euro and dollar bank accounts and enters into forward-purchase contracts for these currencies to mitigate the risk from changing exchange rates.

[19] UK House of Commons, Foreign Affairs Committee, *Refreshing Our Approach? Updating the Integrated Review: Government Response to the Committee's Fifth Report—Fifth Special Report of Session 2022–23*, December 13, 2022.

also outlines plans for defense spending to reach 2.25 percent of GDP by 2025 and an "aspiration" to increase this to 2.5 percent "in the longer term."[20]

Given the MoD's ambitious long-term goals and concurrent requirement to respond to short-term operational pressures, it will need to overcome both internal barriers (e.g., culture, bureaucracy) and the destabilizing impact of a confluence of external trends:

- The UK has experienced an unprecedented period of acute political instability (with three prime ministers in 2022) and faces increased fiscal pressures in the wake of Brexit, the COVID-19 pandemic, and cost-of-living and energy crises.
- The Ukraine crisis has tested the flexibility of the UK's budgetary mechanisms in responding to emerging and unplanned requirements. Aid packages to Ukraine have depleted defense equipment and munitions stockpiles; the UK's support for Ukraine is second only to that of the United States.
- Inflation has been increasing sharply and might force the MoD to cut its budget in real terms. Similarly, the defense sector is highly exposed to foreign exchange rate (i.e., trading) volatility, given the extent of its U.S. imports, primarily aircraft (e.g., F-35Bs, P-8s, AH-64 *Apache* helicopters, CH-47 *Chinook* helicopters).
- The UK's expeditionary focus, global presence, and global commitments (which are similar to those of the United States) require diverse capabilities and equipment for diverse terrain. This requirement sets the UK apart from other medium powers, such as France, Germany, and Japan, with their narrower mission sets.
- Cost growth and escalation challenges have been intensified by the industrial base and supply-chain challenges over the past few years.
- The MoD is under increased pressure to use its budget to boost economic prosperity (through jobs, exports, and so on) and to maximize the environmental and social benefits of spending. New Treasury rules on public procurement give a minimum 10-percent weighting to "social value" in contract award decisions.[21]

Taken together, this mix of long-term and immediate pressures poses significant dilemmas for UK defense planners and those responsible for managing the MoD's finances and executing its spending plans. But the pressures also provide an impetus for ongoing efforts to adapt the MoD's PPBE processes to encourage more agility and innovation, improve value

---

[20] UK Cabinet Office, 2023.

[21] The 2021 *Defence and Security Industrial Strategy* introduced a requirement for all defense contract awards to consider the broader social value of spending, with a minimum of 10-percent weighting in the overall evaluation criteria (MoD, 2021b). This reflects a wider shift in HM Treasury *Green Book* guidance on appraisals and evaluation for public-sector contracts. Social value in this context can include "economic (e.g. employment or apprenticeship/training opportunities), social (e.g. activities that promote cohesive communities) and environmental (e.g. efforts in reducing carbon emissions)" benefits (UK Government Commercial Function, 2020, pp. 2–3).

for money across the portfolio, and enable MoD and the military to deliver increased output despite limited resources.

Relevance to DoD PPBE Reform

Given the UK's significance as a defense actor, DoD could draw lessons from its past and ongoing efforts to promote flexibility, agility, and innovation. Moreover, it is important for U.S. defense leaders to understand the MoD's budgeting cycle because the UK is a critical U.S. ally in terms of global military responsibilities and capabilities, including nuclear weapon capabilities. The UK is a member of the trilateral AUKUS (Australia–UK–United States) security pact, the Combined Joint Expeditionary Force with France, the European Intervention Initiative, the Five Eyes (Australia, Canada, New Zealand, UK, and United States) security agreement, the Five Power Defence Arrangements (with Australia, Malaysia, New Zealand, and Singapore), the Joint Expeditionary Force (which it leads), NATO, and the Northern Group. It is also a veto-wielding permanent member of the United Nations Security Council. Therefore, the MoD interacts frequently and interoperates closely with the U.S. military and intelligence community, and its defense budget and planning decisions are often made in unofficial concert with DoD decisions and priorities.

Although the MoD operates in a different constitutional, political, and fiscal context from its U.S. counterpart, its approach to PPBE could hold insights for DoD. For example, UK government departments are subject to three- to five-year spending reviews, and they are not subject to legislative interference or continuing resolutions. This provides defense planners with a valuable degree of certainty.[22] Multiyear spending reviews make budgeting more rigid than a yearly budget would, but the Treasury and MoD also retain some level of flexibility when translating the medium-term vision of comprehensive spending reviews into annual budgets and plans. The UK also has mechanisms—although imperfect and perhaps not always used as widely as needed—for moving money between accounts and for accessing additional funds in a given fiscal year. As we discuss, these mechanisms include a process known as *virement* for reallocating funds with either Treasury or parliamentary approval, depending on the circumstances. The MoD can make additional funding requests through in-year supplementary estimates sent to parliament. It has other types of flexibility as well, including access to additional Treasury funds to cover UCRs, and it can use the cross-governmental UKISF,[23] which was known as the CSSF until March 2023, or the Deployed Military Activity Pool "to make

---

[22] HM Treasury (also known as the *Exchequer*) acts as both the treasury and finance ministry and owns the Public Spending Framework. It has a statutory responsibility for setting departmental budgets across the government and is internally political to the governing party but not the parliament, ensuring a degree of stability to implement long-term policies (UK subject-matter expert, interview with the authors, October 2022).

[23] UK Cabinet Office, 2023.

available resources to fund the initial and short-term costs of unforeseen military activity," such as responses to natural disasters or support to Ukraine.[24]

Like DoD, the MoD is experimenting with new ways to encourage innovation, including a new dedicated Innovation Fund that allows the chief scientific adviser to pursue higher-risk projects as part of the main R&D budget. The MoD has further supported innovation through incubators, accelerators, and novel contracting practices. However, these strategies have not alleviated some enduring challenges, such as a risk-averse MoD culture and interservice rivalries.

Overview of the UK Government's Budgeting Process

The UK's Public Finance Management Cycle

The MoD provides inputs at various stages of the UK government's public finance management cycle, following Treasury guidance. As mentioned earlier, spending reviews occur every three to five years and provide a high-level view of all government departments' medium-term budgets. To translate those high-level spending plans into the detailed budgets needed to secure parliamentary approval, the annual budget cycle unfolds during the fiscal year, which runs from April 1 to March 31 in the UK. Figure 4.1 shows the major steps in the cycle.

Comprehensive spending reviews set spending limits for future years and guide resource allocations to priority policy areas, something that provides a level of certainty and flexibility. Using these spending limits, for instance, the MoD generates a one-year budget estimate through the annual budget cycle process, along with additional estimates for ten years out (as set out in the MoD's *Defence Equipment Plan*).[25] These outward estimates form the basis for subsequent annual budget cycles.

---

[24] MoD, *Annual Report and Accounts: 2020–21*, January 20, 2022a, p. 22.

[25] The latest unclassified version, which is published annually, covers the period from 2022 to 2032. See MoD, *The Defence Equipment Plan, 2022 to 2032*, 2022c.

FIGURE 4.1

The UK Government's Public Finance Management Cycle

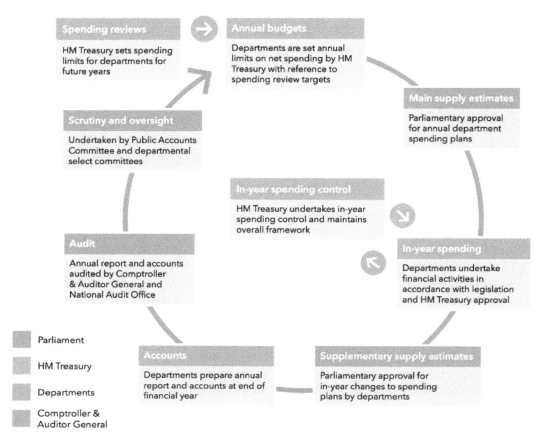

SOURCE: Reproduced from Phillip Trotter, *UK Central Government Public Financial Management System: A Guide for Stakeholders,* Institute of Chartered Accountants in England and Wales (ICAEW), November 2017, p. 4. Reproduced with the kind permission of ICAEW. © ICAEW 2023. No reproduction or re-distribution allowed. ICAEW accepts no responsibility for the quality of the reproduction, or any advice relied there upon.

UK government departments request parliament's authority to spend money through annual appropriations acts. For these requests, departments submit two major types of estimates—main supply estimates (MEs) and supplementary supply estimates (SEs)—to parliament at different times of the year. MEs and SEs are established in close coordination with the Treasury.[26] The ME and SE processes, along with that for excess votes (a procedure for retroactively approving prior-year overruns), unfold as follows:

- MEs
  - These estimates are presented to parliament for approval around the start of the fiscal year. Given the time lag between the start of the fiscal year and the passage of an appropriations act for that year, funding is provided in the early part of the fiscal year through an advance of 45 percent of the previous year's approved spending through a "vote on account." This advance funding ensures the government's continued operation and avoids the possibility of a shutdown.[27]
  - In the case of the MoD, it submits two MEs (MoD Main and the Armed Forces Pensions and Compensation Scheme), typically in April. These MEs are examined by the House of Commons Defence Select Committee and the UK National Audit Office (NAO), an independent public spending watchdog led by the Comptroller and Auditor General. The NAO reports to and supports parliament.[28]
  - Parliament votes separately on the maximum number of personnel permitted to serve in the military in that year (the Votes A convention).[29]
  - For all government departments, parliament approves overall spending limits but does not typically politicize or otherwise dictate the details of exactly how departments spend their money (unlike in the U.S. system).
  - The result is the annual appropriations act, which is typically passed by July and accomplishes three objectives: (1) approving that fiscal year's MEs, (2) appropriating revised sums authorized in the preceding year's SE, and (3) authorizing any excess vote expenditures from the prior year.
- SEs
  - SEs are available so that departments can seek additional resources, capital, or cash during the current fiscal year.
  - These resources can be either substantive (in which case additional resources are required) or token (i.e., no additional resources are required but there is a need to redistribute provisions, authorize higher-than-planned receipts [e.g., from the disposal and sale of old equipment], or approve the use of nonpublic funds).

---

[26] UK subject-matter expert, interview with the authors, November 2022.

[27] UK subject-matter expert, interview with the authors, October 2022.

[28] NAO, "Who We Are," webpage, undated.

[29] UK Parliament, "Erskine May," webpage, undated, Part 5, Chapter 34.

- Building on the ME submitted in April, parliament considers departments' SEs in February of the following calendar year, so shortly before the end of the fiscal year, which runs from April to March.[30]
- excess votes
  - Excess votes occur if a department cannot avoid expenditures beyond the provisions that parliament voted for through the MEs and SEs.
  - Effectively, excess votes retroactively approve overruns from a previous fiscal year because government departments cannot legally spend more money than has been approved by parliament. Therefore, departments are motivated to avoid an excess vote; the House of Commons Public Accounts Committee and other scrutiny functions are highly critical of these situations, and ministers or other senior civil servants may be expected to resign, especially if the excess is significant.

The UK's fiscal and spending framework uses both *accrual-based budgeting* (i.e., budgeting based on when transactions occur rather than when cash receipts or payments are exchanged) and *zero-based budgeting*, in which all activities and programs must be recosted from zero and justified through a set of criteria for prioritizing projects with the highest value for money. The Treasury controls the MoD's spending using the accrual system.[31] The MoD reports its spending on a monthly basis to comply with Treasury reporting requirements.[32] Like those of every other department, the MoD's budget works on a "spend-it-or-lose-it" basis: The money allocated each year must be spent or it is returned without compensation. (Although, as discussed later, there are mechanisms for limited flexibility.[33])

UK and MoD Spending Controls

Once approved, the MoD's spending is subject to a mix of parliamentary and Treasury controls. The levels of control and approval are scaled to the levels of spending and risks involved. But, in general, the most notable controls are departmental expenditure limits (DELs) and annual managed expenditures (AMEs). These controls are further broken down into resource DELs, capital DELs, resource AMEs, and capital AMEs:[34]

- DELs

---

[30] MoD, *Financial Management and Charging Policy Manual, Part 1: Directive*, version 7.0, Joint Service Publication 462, March 2019a, withdrawn November 27, 2020, p. 11.

[31] UK subject-matter experts, interviews with the authors, October and November 2022.

[32] MoD, 2019b, p. 26.

[33] UK subject-matter expert, interview with the authors, October 2022.

[34] MoD, 2019b, p. 3; MoD, 2019a, pp. 2, 5.

- DELs are a department's fiscal limits for discretionary spending. DELs make up most of the MoD's budget. They are planned and budgeted over three- to five-year periods, although per-year spending is voted on annually, as described earlier.[35]
- *Resource DELs* pertain to daily expenses. They are split into cash resource DELs, which cover current expenditures and receipts (including for personnel, equipment support, inventory, infrastructure, and other cash costs), and noncash resource DELs, which cover the depreciation and impairment of property, plants, equipment, and intangibles (and are segregated from other resources or "ring-fenced").
- *Capital DELs* pertain to noncurrent expenditures and receipts, both intangible and tangible, such as "investment in new equipment and infrastructure."[36] Capital DELs are split into fiscal capital DELs, which cover expenditures for equipment that may have civilian uses (e.g., information technology equipment), and single-use military equipment or equipment that has only a military role (e.g., a warship).
- DELs are reported in terms of relevant commodity blocks, such as R&D, infrastructure, service personnel, and civilian personnel.

- AMEs
  - AMEs are "areas of spending that HM Treasury deems unpredictable, difficult to control, and of a size that departments would have difficulty managing within DEL budgets."[37]
  - *Resource AMEs* constitute "mainly provisions, depreciation and impairment and movements in financial estimates."[38]
  - *Capital AMEs* are related primarily to "nuclear provisions," given the sizable share of total MoD expenditures dedicated to the Royal Navy's continuous at-sea deterrent.[39]
  - Given their inherent volatility, AMEs are not planned or budgeted in the same way as DELs but are voted on annually.

Beyond MEs, SEs, excess votes, DELs, and AMEs, which apply to all government departments, defense *operations* are funded separately through the Treasury or, in certain circumstances, the UKISF.[40] Operations are placed into a separate funding category because the costs are likely to be volatile and driven by changing threats and other external events, making them difficult for the MoD to manage within DEL budgets.[41] Operations are subject

---

[35] MoD, 2022d.

[36] UK House of Commons, Defence Select Committee, 2021, p. 2.

[37] Trotter, 2017, p. 15.

[38] UK House of Commons, Defence Select Committee, 2021, p. 2.

[39] UK House of Commons, Defence Select Committee, 2021, p. 2.

[40] As mentioned earlier, UKISF was formerly known as CSSF.

[41] MoD, 2022d.

to their own spending limits—specifically, operations resource DELs (for daily spending) and operations capital DELs (for capital expenditures).

The UKISF was announced in March 2023 and replaces the CSSF, which was launched in 2015 as a unique cross-government fund for tackling cross-cutting issues of peace and stability in fragile countries. It disburses Official Development Assistance (ODA) and other funding to encourage joint work across government departments and agencies and more-agile responses to changing circumstances that are not anticipated in individual departmental budgets.

The UKISF includes the original CSSF's funding for peacekeeping operations (e.g., under the auspices of the United Nations) and management of the Rapid Response Mechanism, a tool for the rapid mobilization of non–ODA funding that has been used to respond to hurricanes in the Caribbean, the Novichok poisonings in Salisbury, and Russia's invasion of Ukraine. Adding to the CSSF's original remit, the UKISF includes funding to sustain the UK's sanction regime and to support national security projects at home, as part of the Economic Deterrence Initiative.[42] The UKISF, like the CSSF before it, is managed by the Joint Funds Unit in the Cabinet Office. This unit funds cross-government programs administrated by the MoD; the Foreign, Commonwealth and Development Office; the Home Office; and other government departments or agencies.[43]

Figure 4.2 provides a breakdown of all types of spending limits that apply to the MoD. The figure distinguishes between (1) the planned and not planned annual defense budget and (2) the defense budget with and without operations costs.

Beyond the various DELs and AMEs, other Treasury controls include the following:

- required approval for any MoD expenditure above £600 million (This is generous compared with the threshold of £50 million used for other departments.)
- monthly and annual reporting requirements; the MoD must provide the Treasury with actual and forecasted spending through the government-wide Online System for Central Accounting and Reporting (OSCAR)
- requirements to explain and justify any major changes in planned costs, receipts, or outputs
- audits by the NAO and the Comptroller and Auditor General.

The MoD's Internal Annual Budget Cycle

To play its role in the UK's public finance management cycle, the MoD splits its internal PPBE-like process into eight top-level budgets (TLBs) corresponding to the eight main MoD organizations and then exerts central oversight to promote jointness across the depart-

---

[42] UK Foreign, Commonwealth and Development Office, "New Fund Announced to Support UK's National Security Priorities," press release, March 13, 2023.

[43] UK Conflict, Stability and Security Fund, "About Us," webpage, undated.

FIGURE 4.2

MoD Departmental Resources, 2021

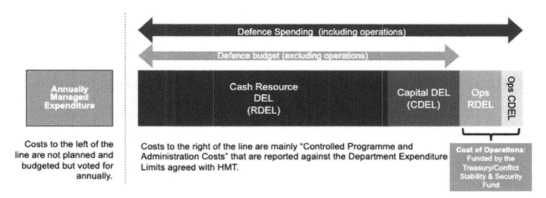

SOURCE: Reproduced from MoD, 2022d (contains public-sector information licensed under the Open Government Licence v3.0).
NOTE: HMT = His Majesty's Treasury; Ops = operations.

ment. The MoD uses the centralized Planning, Budgeting, and Forecasting (PB&F) system to inform budget decisions and implementation. According to Joint Service Publication (JSP) 462, which serves as the MoD's manual for financial management, PB&F provides the "'single version of the truth' at every stage of the programming and budgeting process."[44] (There are plans to make PB&F a cloud-based application, reducing the need for some teams to maintain various Microsoft Excel spreadsheets and engage in other recordkeeping in parallel with their PB&F reporting.)[45]

Using the multiyear comprehensive spending reviews and associated integrated or defense reviews,[46] UK defense leaders consider which missions or high-level concepts should be developed into tangible programs. The MoD next issues its Command Paper and associated departmental plans, translating high-level requirements into tangible programs. The implementation of these decisions is laid out in each year's Defence Equipment Plan, with estimated budgets and timelines for the individual services in both the year 1 annual budget cycle and outward for ten years.

The MoD sets annual TLBs for its eight main organizations: the Royal Navy, British Army, Royal Air Force, UK Strategic Command, MoD Head Office, Defence Equipment and Support (DE&S), Defence Infrastructure Organisation, and Defence Nuclear Organisation. The MoD negotiates with these organizations through "demand signals,"[47] leaving it up to them to program against those required outputs and effects, and then feeds the proposed programs

---

[44] MoD, 2019b, p. 23.

[45] UK subject-matter expert, interview with the authors, November 2022.

[46] UK subject-matter expert, interview with the authors, November 2022.

[47] UK subject-matter expert, interview with the authors, October 2022.

into a centralized prioritization process.[48] As JSP 462 states, "Within a fixed budget set by the Spending Review/Round process, the [annual budget cycle] is therefore essentially a process of prioritization, with decisions to allocate more resources to areas of high priority requiring compensating savings elsewhere in the Defence Programme."[49]

Features of the MoD's Internal Budgeting Processes

In this section, we explore the internal workings of the MoD's PB&F system, focusing on key actors, processes, and outputs.

Decisionmakers and Stakeholders

UK Defence, the name given to the combined MoD and military services, is structured around the MoD Head Office and a complex ecosystem of delivery and enabling organizations. The most powerful are the Front Line Commands (FLCs). The FLCs consist of the services—specifically, the Royal Navy (by convention, the senior service), the British Army, and the Royal Air Force—plus the joint UK Strategic Command. Formerly known as Joint Forces Command, UK Strategic Command is the newest FLC and is responsible for multidomain integration; joint, typically niche capabilities (such as cyber, space, and special forces); an array of central functions, such as joint concepts and doctrine development; and joint professional military education.

The four FLCs fall under the political leadership of the Secretary of State for Defence (an elected member of the House of Commons, as of this writing, Ben Wallace MP, a former soldier), who is supported by four junior ministers: the Minister of State for the Armed Forces; the Minister for Defence Procurement; the Minister for Defence People, Veterans and Services Families; and the appointed Minister of State in the House of Lords, who represents the government on defense-related business in the upper chamber. Other than a small number of political special advisers, ministers are supported by senior civil servants who are not political appointees and, therefore, are supposed to remain independent of party politics and execute the mandate of the government of the day. Ministers are also supported by the four-star military chiefs, the Chief of the Defence Staff, the Vice Chief of the Defence Staff, and the chiefs of the FLCs. Figure 4.3 shows the MoD organization chart.

The various UK Defence teams and functions are coordinated through the Defence Operating Model, the latest version of which was published in September 2020. As shown in Figure 4.4, stakeholders in this operating model are mapped against different phases of an overarching activity cycle: direct, enable, acquire, generate and develop, and operate.

---

[48] UK subject-matter expert, interview with the authors, October 2022.

[49] MoD, 2019b, p. 17.

FIGURE 4.3
MoD Organization Chart

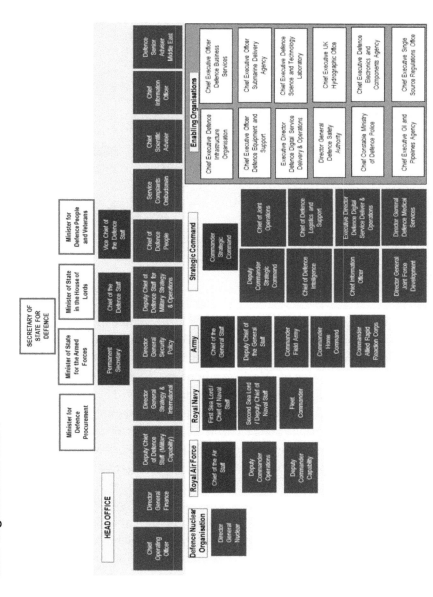

SOURCE: Reproduced from MoD, *How Defence Works*, version 6.0, September 2020a, p. 10 (contains public-sector information licensed under the Open Government Licence v3.0).

FIGURE 4.4

Defence Operating Model

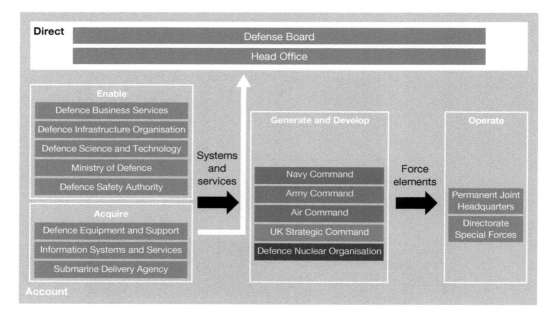

SOURCE: Features information from MoD, 2020a.
NOTE: Permanent Joint Headquarters and Directorate Special Forces are within UK Strategic Command.

The FLCs and the Defence Nuclear Organisation (in the middle of Figure 4.4) annually file planning and programming options through the PB&F as part of their responsibilities to generate and develop relevant defense capabilities and forces. The MoD Head Office (strategic headquarters) then decides how to prioritize among these options and directs the FLCs and other delivery organizations to execute the agreed-on plans. The civilian Secretary of State for Defence and other civil service counterparts have final approval.

As shown in the model, various actors enable the business of defense. For example, Defence Science and Technology is responsible for R&D and the work of the Defence Science and Technology Laboratory, while the Defence Infrastructure Organisation is responsible for recapitalizing and maintaining installations and other infrastructure.

Other actors are responsible for acquiring and supporting the products and services needed by the FLCs and the Defence Nuclear Organisation. Most notable is DE&S, headquartered at Abbey Wood. DE&S is a trading entity staffed by a mix of civil servants, military officials, and contractors, who together number 11,500 staff across 150 sites, handle most procurement and life-cycle support for the MoD, and execute the ten-year Defence Equipment Plan.[50] Smaller, specialized agencies are also active in procurement, including Defence Digi-

---

[50] MoD, Defence Equipment and Support, "Who We Are," webpage, undated. There have been debates about whether DE&S should move to a government-owned, contractor-operated model instead of being a

tal (formerly Information Systems and Services). Part of UK Strategic Command, Defence Digital reports to the MoD's chief information officer and is responsible not only for digital strategy and policy but also for acquiring and supporting information technology across both the corporate and military elements of UK Defence.[51] Another key specialized actor is the Submarine Delivery Agency (SDA), an executive agency carved from DE&S in 2018 and given the flexibility to direct additional focus to high-cost, high-risk programs associated with recapitalizing and maintaining the UK's continuous at-sea deterrent.[52] (In launching the SDA, the MoD learned from the successful model of the arms-length Olympic Delivery Agency that supported planning and budgeting for the 2012 London Olympics.[53])

Collectively, the Defence Operating Model links strategic and policy objectives (direct) to force and capability development priorities (generate and develop), which inform spending by enabling organizations (enable) and the acquisition, upgrade, and maintenance of relevant equipment and services (acquire)—all as needed to meet FLC and Defence Nuclear Organisation requirements. The various defense lines of development (DLODs) are then integrated and put through a force development and generation cycle to deliver force operational capabilities.[54] Operations are then overseen by Permanent Joint Headquarters (operate), which reports to MoD leadership.

Through the PB&F process, the MoD also interacts with external actors, including the Treasury, parliament, and NAO, as well as (to a more limited extent) industry. The Defence and Security Industrial Strategy (DSIS), published in 2021 alongside the Integrated Review and Defence Command Paper, signaled a desire to deepen the MoD's relationship with industry, balancing open competition in some areas with long-term strategic partnerships in others.[55] To this end, the DSIS outlined plans for strengthening public-private partnerships,

---

bespoke central government trading entity. However, a competition to award a long-term contract to industry collapsed in 2013 when the bidding consortia pulled out. See Robin Johnson, "UK: Ministry of Defence (MoD) Proposals for GOCO Shelved in Favour of DE&S Plus Variant," Eversheds Sutherland International, December 11, 2013.

[51] MoD, "Defence Digital," webpage, undated-b.

[52] MoD, Submarine Delivery Agency, "About Us," webpage, undated.

[53] Richard Johnstone, "First Class Delivery: What the MoD Team Renewing the UK's Nuclear Submarines Learnt from the Olympics and Crossrail," *Civil Service World*, May 29, 2018.

[54] The DLODs are the UK equivalent to the U.S. DOTMLPF framework and comprise training, equipment, personnel, information, concepts and doctrine, organization, infrastructure, logistics and interoperability (often referenced by the mnemonic TEPID OIL).

[55] MoD, 2021b. For example, the MoD might coordinate with industry where there are monopsony-monopoly dynamics at play in the UK market, as there are for nuclear submarine production, or where multiannual settlements and joint capability planning would be particularly useful, as with the ten-year Portfolio Management Agreement and the Team Complex Weapons program, a collaboration between the MoD and the UK's main missile manufacturer. See MBDA Missile Systems, "The Portfolio Management Agreement," webpage, undated.

improving supply chains, and "developing the Joint Economic Data Hub" as part of an effort to further improve decisionmaking around defense budgets, plans, and programs.[56]

To communicate its intentions to industry, the MoD outlines its ten-year estimates and associated procurement priorities through the Defence Equipment Plan, which is updated and published annually. This plan is intended to send a demand signal to guide industry investments in production facilities, skill development, and R&D. The MoD also works closely with a more select group of large defense companies that are "capable of managing the complex financial, technological and engineering demands of delivering highly complex systems, with [small- to medium-sized enterprises] typically engaged in their supply chains."[57] In the DSIS, the MoD described its desired relations with industry as a "virtuous circle," whereby the MoD's R&D or procurement funding aids the private sector in developing new technologies that will then be used by the MoD and exported abroad for the overall benefit of the UK.[58]

For-profit contractors and nonprofit research institutions also support the activities of the MoD's Financial and Military Capability team, the TLBs, and such delivery organizations as DE&S across the life cycle of the PB&F process. These contractors support analyses of alternatives, red team draft plans, operating concepts, and commercial strategies, and they offer advice, expertise, and tools for cost estimation and cost modeling. This support often goes through the Cost Assurance and Analysis Service (CAAS) within DE&S and various management consultancy framework contracts.[59]

Planning and Programming

To align the complex ecosystem of the Defence Operating Model with the annual budget cycle, PB&F decisionmaking is somewhat decentralized. This decentralization builds on the Levene Reforms of the early 2010s, which were named after Lord Peter Levene. These reforms sought to bring about the following improvements:[60]

- streamlining decisionmaking
- empowering the services and the other enabling and delivery organizations to take control of their own budgets and "advise on the best balance between manpower, training, equipment and support . . . that are needed to deliver the Defence requirement" rather than having these solutions dictated from the top down

---

[56] MoD, 2021b, p. 8. Examples of such efforts include better understanding the skills base among major suppliers and, thus, their capacity to deliver or better understanding the broader spillover benefits of defense spending to regional or national prosperity and, thus, its overall returns beyond security.

[57] MoD, 2021b, p. 15. Also see Bill Kincaid, *Changing the Dinosaur's Spots: The Battle to Reform UK Defence Acquisition*, Royal United Services Institute, 2008, p. 108.

[58] MoD, 2021b, p. 5.

[59] This has included work for CAAS by RAND Europe, part of the RAND Corporation.

[60] MoD, "Defence Secretary Unveils Blueprint for Defence Reform," June 27, 2011.

- offsetting and balancing the risk of incoherence as a result of this decentralization by enhancing the institutional and leadership focus on jointness and integration (including through the creation of Joint Forces Command, now UK Strategic Command)
- increasing the focus on affordability and budget discipline at all levels.

Within its overall budget, as outlined in comprehensive spending reviews and approved annually by parliament, the MoD sets annual TLBs for its eight main organizations, which then bid for money in response to demand signals (i.e., required outputs and outcomes). The MoD provides the central analysis, planning, and programming functions and negotiates with the TLB organizations (and Treasury) to determine how best to allocate finite resources across its portfolio.[61] Each of the service components forecasts a baseline estimate, to which the MoD typically adds 12–15 percent to ensure that no organization overspends.

Where choices need to be made among competing priorities, it is the initial responsibility of each TLB organization to interact with other stakeholders to determine which option is most suitable.[62] Although the TLB organizations should attempt to deconflict their bids, it is the responsibility of the centralized functions in the MoD to make the ultimate decisions about how best to prioritize among the various TLB proposals, given the available inputs and required outputs and outcomes.

Within the MoD, the programming and budgeting processes are owned by the Director General, Finance, who works closely with the Deputy Chief of the Defence Staff for Military Capability, a member of the MoD's Financial and Military Capability team. They are supported by the Director of Financial Planning and Scrutiny and the Assistant Chief of the Defence Staff for Capability and Force Design, also a member of the Financial and Military Capability team. Day-to-day process responsibilities are delegated to the Head of Defence Resources, whose team provides instructions and guidance on the annual budget cycle timetable and process support to the TLB organizations.

The central teams in the MoD coordinate on various aspects of strategy development and planning, financial processes, performance and reporting processes, and capability and force development processes, as shown in Figure 4.5. These teams translate the Integrated Review, Command Paper, and other policies to a set of planning documents at different levels of granularity, including the overall defense strategy, the MoD's annual defense plan, any subordinate strategies (e.g., by DLOD or function), and command plans or corporate plans at the TLB level. These documents inform and are informed by the annual budget cycle and the meeting cycles of various decisionmaking, performance, and risk-review bodies, such as the Executive Committee and Defence Board.

When it comes to making decisions about priorities across the TLBs, the central MoD teams draw on a variety of data sources (including the PB&F system) and analyses. The MoD recently introduced machine-learning tools to automate some PB&F estimates and is shift-

---

[61] UK subject-matter expert, interview with the authors, October 2022.

[62] MoD, 2019b, pp. 20–22.

FIGURE 4.5

MoD Planning and Performance Reporting Processes

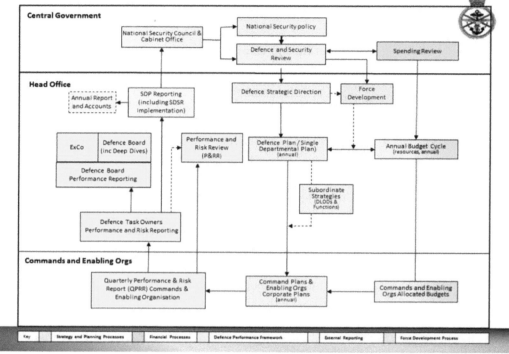

SOURCE: Reproduced from MoD, 2020a, p. 15 (contains public-sector information licensed under the Open Government Licence v3.0).

NOTE: Inc = including; Orgs = organizations; SDP = Single Departmental Plan; SDSR = Strategic Defence and Security Review. Green shading indicates strategy and planning processes, orange shading indicates financial processes, yellow shading indicates the defence performance framework, blue shading indicates external reporting, and purple shading indicates the force development process.

ing to a cloud-based solution, which might be adopted government-wide.[63] Other techniques are more established. Because few MoD activities produce monetizable benefits, there is relatively limited use of cost-benefit analysis. Instead, most appraisals take the form of a cost-effectiveness analysis, including a specialized variant known as a *combined operational effectiveness and investment appraisal*, which is used for considering how to maximize value for money with new military equipment.[64]

---

[63] UK subject-matter expert, interview with the authors, October 2022.

[64] This involves an estimate of the total life-cycle costs of options to meet a particular requirement in the Investment Appraisal, the identification of "individual parameters contributing to overall performance," and an assessment of each option against these parameters as part of the Operational Effectiveness Appraisal before the two separate assessments are combined to provide an overall picture of the cost-

To ensure that they select the right methodological approach and execute it successfully, MoD teams can draw on guidance in various MoD and Treasury documents, including the following:

- JSP 462, *Financial Management and Charging Policy Manual*, is the primary document that the MoD references when planning, programming, and budgeting programs. It provides reasonably high-level guidance for planning and executing the annual budget. This manual is revised at least every two years as practices evolve.[65]
- JSP 507, *Investment Appraisal and Evaluation*, details how to undertake appraisal and evaluation within the MoD and supplements the Treasury's *Green Book* by interpreting and applying it to specific defense challenges (e.g., difficulties with cost estimation, the lack of clearly monetizable benefits).[66]
- The *Green Book* is the overarching policy and practical guidance for investment appraisals and evaluation across the UK government.[67] It is produced and maintained by the Treasury and is complemented by supplementary guidance in the *Aqua Book* (on producing quality analysis), *Magenta Book* (on designing evaluations), and *Orange Book* (on risk reporting and management), as well as the *Public Value Framework*. The latter elaborates on the concept of measuring value for money and the translation of funds into policy outcomes in the public sector, including for defense, where traditional measures may not always be appropriate.[68]

Central teams, such as the MoD's Financial and Military Capability, can also draw on data, expertise, and analysis from science and technology, domain, capability, or wargaming specialists in the Defence Science and Technology Laboratory; futures, concepts, and doctrine specialists in the Development, Concepts, and Doctrine Centre; externally contracted support; or multiple layers of cost data and modeling at the annual budget cycle, TLB, delivery agency, or individual project levels. For example, DE&S has a project control function that includes subteams specializing in cost control, scheduling and planning, estimating, risk management, and

---

effectiveness of each option. See MoD, *Investment Appraisal and Evaluation, Part 1: Directive*, version 6.0, Joint Service Publication 507, January 2014, p. 4.

[65] MoD, 2019a; MoD, 2019b.

[66] MoD, 2014, p. 3.

[67] HM Treasury, 2022b.

[68] HM Treasury, *The Aqua Book: Guidance on Producing Quality Analysis for Government*, March 2015; HM Treasury, *Magenta Book: Central Government Guidance on Evaluation*, March 2020; Government Risk Protection and the Risk Centre of Excellence, *The Orange Book: Management of Risk—Principles and Concepts*, 2020; HM Treasury, *The Public Value Framework: With Supplementary Guidance*, March 2019. For more on the Public Value Framework, see James Black, Richard Flint, Ruth Harris, Katerina Galai, Pauline Paillé, Fiona Quimbre, and Jessica Plumridge, *Understanding the Value of Defence: Towards a Defence Value Proposition for the UK*, RAND Corporation, RR-A529-1, 2021.

broader coordination and management of information and decisionmaking.[69] The MoD also has an independent "centre of excellence for pricing and costing support" (the aforementioned CAAS), which was first established around World War I.[70] In FY 2020–2021, CAAS executed "89 independent cost estimate reviews" to determine relative risk.[71]

Other central MoD teams regulate, advise on, or support spending decisions and execution in specific contexts. For example, the Single Source Regulations Office (SSRO), an executive nondepartmental public body, aims to ensure that the MoD receives value for money and that prices are fair and reasonable in defense contracts awarded without competition, in line with the Defence Reform Act 2014 and Single Source Contract Regulations 2014.[72] Still, in its annual report for 2022, the SSRO remarked on continuing shortfalls in the quality and timeliness of contract and supplier reports—a statutory requirement that enables the SSRO's pricing and compliance reviews—indicating that this kind of regulation remains an area for further improvement and/or legislative reform.[73]

In 2020, the MoD also introduced a new Evaluation Team, intended as another center of excellence, that it hopes to maintain and expand. This team examines projects underway to glean "better corporate and project level understanding of what works in project delivery, reducing risk and uncertainty and helping the Department stop avoidable costly mistakes."[74]

Combining various analytical techniques and sources of evidence allows the MoD to compare alternatives and balance its investments and force design and force mix decisions in pursuit of an optimal allocation of resources across the defense portfolio. In practice, of course, there are still substantial interservice rivalries at play. For example, there is a perception that the Navy and Air Force have benefited the most from defense reviews and equipment plans over the past decade, while the Army has achieved less success in pushing its case within the MoD or in resisting cuts to troop numbers and procurement budgets.[75] Others argue, however, that this result reflects not merely the UK's inevitable focus on certain domains as an island nation with an interest in global power projection but also a virtuous feedback loop between the UK's poor record of executing major land equipment programs in recent decades (e.g., Challenger 2 and Warrior upgrades or the acquisition of Boxer and Ajax)[76] and the

---

[69] Tim Sheldon, "Establishing a Project Controls Function at the UK Defence Equipment and Support Organisation," presentation at the Project Controls Expo 2017, London, November 16, 2017.

[70] MoD, "Ministry of Defence Commercial," webpage, updated December 12, 2012.

[71] MoD, *The Defence Equipment Plan, 2021–2031*, 2022b, p. 20.

[72] UK Single Source Regulations Office, "About Us," webpage, undated.

[73] UK Single Source Regulations Office, *Annual Compliance Report 2022*, November 2022.

[74] MoD, 2022b, p. 27.

[75] Michael Clarke, "Army, Navy and RAF: Winners and Losers of Defence's Transformation," webpage, Forces.net, March 22, 2021.

[76] The Public Accounts Committee has been especially critical of the Army and MoD's program to acquire the Ajax armored vehicle, which was built by General Dynamics UK. That program has been beset by

Army's struggles in convincing the MoD and parliament that its proposals represent credible and affordable solutions.

The MoD faces other enduring challenges in planning and programming, such as how best to manage the complex dependencies between its TLB organizations. For example, the Army might be hindered by changes to an Air Force budget that delay an air capability's entry into service.[77] This interdependency is reflected in the MoD's recognition of the need for further improvements in organizational culture, processes, skills, and use of decision support aids if the MoD is to realize its ambitions to go beyond jointness and achieve its vision of multi-domain integration (the UK analog to the U.S. joint all-domain operations concept) and thereby to maximize its ability to compete in a quickly deteriorating threat environment.[78]

Budgeting and Execution

As discussed earlier, budgets are broken down by commodity blocks (e.g., capital DELs, resource DELs, AMEs) and by activity (e.g., personnel). Figure 4.6 provides a high-level breakdown of these MoD departmental resources for 2021.

Reallocating funds midyear might be discouraged or difficult, but in theory, there are several formal mechanisms for doing so. First, an estimate of already voted funds can be added to or moved within TLB programs with Treasury approval, provided they stay within the same commodity block (i.e., resource DELs, capital DELs, AMEs). Additional funding can also be requested from parliament for one or more TLB programs as an SE. A substantive SE occurs when a new service cannot be funded "from savings within the DEL," when there is a "gross overspend or a forecast shortfall of Appropriation in Aid," when "authority is needed to meet an excess on one DEL either from savings on another DEL or from drawing on any Budget Flexibility," or when "there is to be a transfer of provision to or from another Department."[79] A token SE occurs when there is a need for a new capability that is not part of a DEL but "could be met from savings within the DEL," when "existing Estimates provision is to redistributed within the DEL," or when "the cost of a new service is to be met from non-public funds."[80]

MoD funds can also be directly transferred between programs within a DEL or AME in a process known as virement. There are numerous restrictions on virements, however. The

---

problems and delays. See UK House of Commons, Public Accounts Committee, *Seventh Report of Session 2022–23: Armoured Vehicles: The Ajax Programme*, May 25, 2022.

[77] Lucia Retter, Julia Muravska, Ben Williams, and James Black, *Persistent Challenges in UK Defence Acquisition*, RAND Corporation, RR-A1174-1, 2021.

[78] MoD, *Integrated Operating Concept*, August 2021c; MoD, *Multi-Domain Integration*, Joint Concept Note 1/20, November 2020b.

[79] MoD, 2019a, p. 7.

[80] MoD, 2019a, p. 7.

FIGURE 4.6
MoD Departmental Resources, 2021

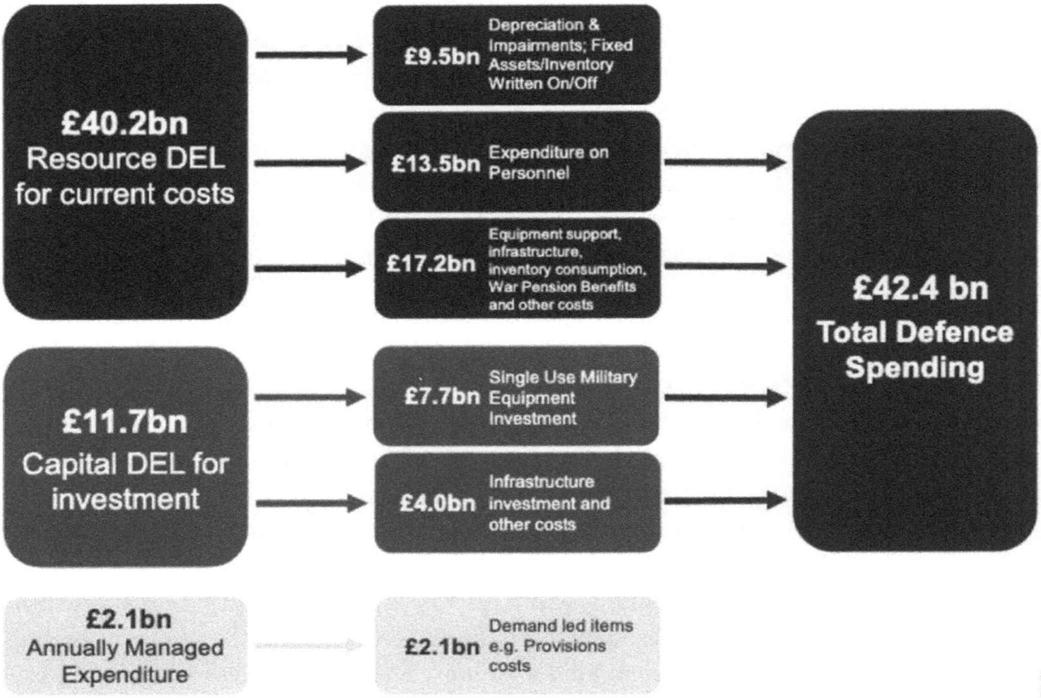

SOURCE: Reproduced from MoD, 2022d (contains public-sector information licensed under the Open Government Licence v3.0).

MoD cannot, for instance, transfer funds from a "non-voted provision to [a] voted provision." It also "cannot vire between voted budgetary provision[s]," such as between a resource DEL and a capital DEL.[81] Such changes would require an SE. Meanwhile, virement from programs (such as acquisition of equipment) to administrative functions (such as personnel management) requires Treasury approval, whereas virement from administration to programs *is* allowed without Treasury approval to encourage efficiency and savings in favor of FLCs. Virement from ring-fenced to non–ring-fenced accounts (e.g., from noncash to cash resource DELs) is not usually permitted.[82]

The Treasury will typically escalate proposed changes to a parliamentary vote on an SE when the expenditure is novel or contentious, reflects a major policy change, or could involve heavy future liabilities. In other instances, the MoD may seek an SE if it underspends: It can ask to carry forward some of the unspent funds to the next year through a parliamentary

---

[81] MoD, 2019a, p. 6.

[82] MoD, 2019a, p. 6.

vote, but if it does not, or if the request is rejected, the MoD must use it or lose it within the fiscal year.

The most volatile elements of the MoD's budget are not DELs; instead, they are treated as AMEs or, in the case of operations, covered by the Treasury directly or by the UKISF.[83] Therefore, the MoD can leverage additional flexibility by requesting additional funds from the Treasury to address urgent and unanticipated needs, as it did to support Ukraine. Such requests build on practices refined during the wars in Iraq and Afghanistan, which prompted a series of short-notice requirements for the rapid development and acquisition of new capabilities. Those wars led to an evolution in the processes for fulfilling UCRs, which were previously known as urgent operational requirements. According to the Defence and Security Public Contracts Regulations, UCRs are

> requirements for military or sensitive security goods arising from:
>
> - a need to rapidly respond to an unforeseen threat or mission critical operational risk or to address a new or unforeseen essential safety requirement that poses a risk to life
> - an urgent need to procure a unique requirement that cannot be met through an in-service capability or any other mitigation measures.[84]

UCRs allow the MoD to tackle capability shortfalls that cannot be met within the timeline of the normal acquisition cycle. That being said, "value for money considerations still apply to UCRs although the weighting given to time, performance and cost dynamics may not be equal."[85]

Beyond supporting the MoD, the Treasury has also developed innovative methods of funding rapid cross-governmental responses to crises, such as the COVID-19 pandemic. During the pandemic, the government adapted the rules governing the Contingencies Fund, which the Treasury uses to fulfill funding requests prior to parliamentary approval, meet a temporary department need, or make up a cash deficiency. Given the speed and scale of changes in the demand on public services because of the COVID-19 pandemic, as well as the risk of delay in parliamentary approval processes, the Treasury expanded the fund to offer the government more flexibility. Whereas the fund's capital is typically limited to 2 percent of the government's authorized supply expenditure for the previous fiscal year, this limit was temporarily increased to 50 percent for FY 2020–2021 before it was tapered to 12 percent for FY 2021–2022.[86]

---

[83] As discussed earlier, the UKISF was formerly known as CSSF.

[84] MoD, "DSPCR Chapter 9: Procuring Urgent Capability Requirements (UCRs)," webpage, updated September 27, 2023.

[85] MoD, "The Defence and Security Public Contracts Regulations (DSPCR) 2011," webpage, updated November 28, 2022e.

[86] These changes were the result of the Contingencies Fund Acts of 2020 and 2021, respectively. See HM Treasury, "Contingencies Fund Account 2021 to 2022," webpage, June 16, 2022a.

Innovation

The 2021 DSIS called for closer collaboration between the MoD and industry to develop high-tech military and dual-use (defense and civil) programs.[87] The MoD's budget allocates a substantial amount to R&D (set to reach £6.6 billion over four years and starting with more than 10 percent of the capital DEL for FY 2022–2023).[88] This represents a reversal in historically downward trends in R&D expenditure since the end of the Cold War, as shown in Figure 4.7.

Sources suggest that, if the UK intends to become a "Science and Tech superpower," as mandated in the 2021 Integrated Review, the MoD will need to play an important role.[89] The MoD's chief scientific adviser also has the discretion to use the ring-fenced Innovation Fund. Sources described this added flexibility as "work[ing] really well," despite the fund's relatively small size.[90]

FIGURE 4.7

MoD Research and Development Net Expenditures, 2003–2022

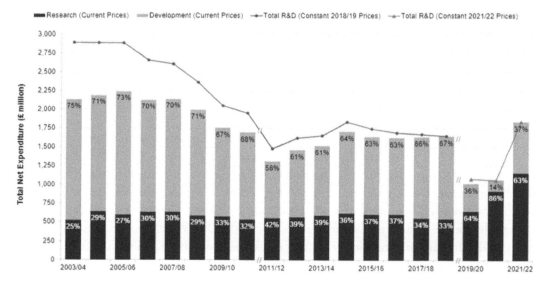

SOURCE: Reproduced from MoD, 2022d (contains public-sector information licensed under the Open Government Licence v3.0).

---

[87] MoD, 2021b, p. 9; UK subject-matter expert, interview with the authors, November 2022.

[88] MoD, 2022b, p. 11.

[89] UK Cabinet Office, 2021, p. 4. See also Paola Fusaro, Nicolas Jouan, Lucia Retter, and Benedict Wilkinson, *Science and Technology as a Tool of Power: An Appraisal*, RAND Corporation, PE-A2391-1, November 2022; UK subject-matter expert, interview with the authors, October 2022; MoD, "Defence and Security Accelerator," webpage, undated-c.

[90] UK subject-matter expert, interview with the authors, October 2022; MoD, undated-c.

Besides its main research efforts through the Defence Science and Technology Laboratory, the MoD also operates the Defence and Security Accelerator, a program designed to fund small- and medium-sized enterprises to conduct fast-tracked R&D. Each FLC operates its own innovation hub.[91] Despite these developments, some believe that the MoD has generally not considered in-house R&D to be a crucial element of its mission, given the pressure on budgets (according to an interviewee), and the innovation burden has often been pushed to the private sector.[92]

Oversight

The Treasury normally submits the MoD's ME to parliament in April. Parliament, especially the House of Commons Defence Select Committee and the NAO, then scrutinize this estimate.[93] The ME submission to parliament is made up of three parts:

- **Part I** is primarily composed of the *ambit*, a detailed description of what "Resource, Capital, Cash, and Non-Cash items" will be consumed that year.[94] Spending beyond what is described in the ambit is not permitted.
- **Part II** provides a detailed breakdown of projected spending by service component. The total is the Treasury settlement that appears as the true budget in OSCAR (the Treasury's system), which is distinct from the MoD's internal PB&F system. Part II also includes income reporting (which can be used to reduce the amount that the MoD requests from parliament), current resource expenditures, capital expenditures, capital income, and key parliamentary controls.[95]
- **Part III** details extra receipts, or unspent funds that would ordinarily need to be returned. However, departments can keep up to 20 percent of resource and capital DEL funds in excess of what was in the Treasury settlement, provided they are backed by supplementary provisions of SEs.[96] The rest must be returned to the general consolidated fund for Treasury disbursement across departments and ministries the following fiscal year. Part III also includes a forecast statement of comprehensive net expenditures, doc-

---

[91] UK subject-matter expert, interview with the authors, November 2022. Hub examples include NavyX, J-Hub, and the Air Force's Rapid Capabilities Office, which is closely modeled on its U.S. namesake. The Rapid Capabilities Office is supported by Will Roper, a former Pentagon official. After his time as Assistant Secretary of the Air Force for Acquisition, Technology, and Logistics, Roper was made an auxiliary honorary group captain in the Royal Air Force and advised on the service in standing up its Rapid Capabilities Office. See Valerie Insinna, "Former US Air Force Acquisition Czar Could Help the UK Build Its Future Fighter," *Defense News*, September 14, 2021.

[92] UK subject-matter expert, interview with the authors, November 2022.

[93] UK subject-matter expert, interview with the authors, October 2022.

[94] MoD, 2019b, p. 2.

[95] MoD, 2019b, p. 2.

[96] MoD, 2019a, p. 7.

umentation showing a reconciliation of resource expenditures between estimates and budgets, and other supporting information.[97]

Because the government commands a working majority of the House of Commons, MPs do not interfere excessively in the specifics of these estimates. That is, unlike Congress, MPs would not insist on a particular capability or basing strategy, for example. Instead, decisions about major defense programs are part of comprehensive spending reviews and defense strategic reviews, which normally occur every three to five years. This time horizon means that, in theory, it might be five years before a new major defense program is reviewed by the government of the day. However, changes can be made outside these cycles, even mid–fiscal year, and a substantive SE can be submitted to parliament if a particular system or technology is needed more urgently.[98]

Once funding has been approved, there are specific legal rules for defense spending.[99] An individual program must spend what is allocated to it for the fiscal year and not exceed its spending cap without permission. More broadly, all expenditures require parliamentary approval of the TLB, expenditures that parliament approves for one military service cannot be spent by another, funding beyond what was approved requires an SE, overspending requires explicit permission, and parliament must approve any emergency funding.[100] As mentioned earlier, an excess vote may be required to approve any overruns from previous fiscal years to ensure that the government does not break these rules.

The MoD and external stakeholders use several mechanisms to ensure that their spending matches what has been approved by parliament. Internally, all expenses from the service components and related organizations are tracked in the MoD's PB&F system. Service components and associated organizations are required to "submit Reports setting out their overall position in terms of delivering the agreed forward programme and the key issues and risks."[101] The Defence Equipment Plan provides annual program updates. Externally, the Treasury's Head of Defence Resources oversees defense spending through OSCAR. The MoD must submit to the Treasury, through OSCAR, in-year management costs for TLBs and other trading entities sitting outside the commodity blocks by the ninth day of every month to ensure a continued flow of cash.[102]

The NAO, the Comptroller and Auditor General, and House of Commons Public Accounts Committee audit the MoD's ME (and any SEs) annually to ensure (1) transparency and taxpayer trust, (2) that spending matches what was approved by parliament, and (3) value

---

[97] MoD, 2019b, p. 3.

[98] MoD, 2019a, p. 11.

[99] MoD, 2019a, p. 2.

[100] MoD, 2019a, p. 2.

[101] MoD, 2019b, p. 23.

[102] MoD, 2022d; MoD, 2019b, pp. 25–26.

for money.[103] According to JSP 462, the NAO audit focuses on the three Es: economy, efficiency, and effectiveness. The NAO is also concerned with value for money.[104] Once the NAO audit has been published, the House of Commons Public Accounts Committee reviews it. If the audit results in modified opinions, the MoD can respond to the audit with a document known as a *treasury minute*. The MoD's permanent undersecretary (the most senior civil servant in the department) would then appear before the Public Accounts Committee to update its members on Treasury minutes and detail how the MoD would respond to the committee and NAO reviews. The MoD must continue to do so until all concerns are addressed.[105] Within the MoD, project controls and evaluation teams similarly undertake internal reviews of individual programs and projects to identify risks and other issues.

Although an advisory agency like the NAO has no explicit political power, the need for transparency is a powerful tool to exert healthy pressure on defense leaders and ensure adequate budgeting of their missions, as stated in the strategy.[106] Furthermore, the NAO conducts deep-dive analyses of specific programs to identify risk factors, lessons, and good practices for future programs. The NAO undertakes or commissions other types of analyses and guidance documentation to build both parliamentary and cross-governmental knowledge on effective program or project design, execution, and evaluation.[107] As one example, a 2021 NAO report examined 20 major MoD programs with a combined life-cycle cost of more than £120 billion to identify recurring insights and lessons for improving the performance of major equipment contracts.[108]

The SSRO also plays a narrow but important role in overseeing projects that involve single or sole-source contracts. There are major examples of such contracts in the MoD, given the monopoly-monopsony dynamics of some portions of the UK defense sector. For example, there is only one credible domestic prime contractor—BAE Systems—for such capabilities as aircraft carriers, nuclear attack and missile submarines, and multirole combat aircraft. As of 2022, the SSRO and its Defence Contract Analysis and Reporting System (DefCARS) database collected and analyzed data on more than 440 contracts estimated to be worth more than £66 billion.[109]

---

[103] MoD, 2019b, pp. 6–8; UK subject-matter experts, interviews with the authors, November 2022.

[104] MoD, 2019b, p. 6.

[105] MoD, 2019b, p. 8.

[106] UK subject-matter expert, interview with the authors, November 2022.

[107] These analyses include cold reviews and reports by RAND Europe researchers. The NAO similarly organizes secondments to build staff knowledge of good practices in various sectors. One area in which the NAO has offered recurring guidance is agile delivery. See NAO, *Governance for Agile Delivery: Examples from the Private Sector*, July 2012; NAO, *Use of Agile in Large-Scale Digital Change Programmes: A Good Practice Guide for Audit and Risk Assurance Committees*, October 2022a.

[108] Gareth Davies, *Improving the Performance of Major Equipment Contracts*, National Audit Office, June 22, 2021.

[109] UK Single Source Regulations Office, 2022, p. 4.

Yet another layer of scrutiny and assurance applies to the largest cross-governmental projects, and there are many in the MoD's portfolio. In January 2016, Infrastructure UK and the Major Projects Authority merged to form the Infrastructure and Projects Authority (IPA).[110] Under a mandate from the prime minister, the IPA collaborates with the Cabinet Office, the Treasury, and other departments and serves as a center of excellence for delivery and risk management of major projects. It also offers additional oversight, support, and assurance to such projects across the UK government.[111] In this context, *major projects* are defined as those that "require spending over and above departmental expenditure limits, require primary legislation, [and/or] are innovative or contentious."[112]

Collectively, the 235 projects in the Government Major Projects Portfolio (GMPP) have a life-cycle cost of around £678 billion and offer monetized benefits of £726 billion. Fifty-two are MoD projects, with around £195 billion in total costs. They include 45 major projects focused on developing new military capabilities corresponding to £174 billion in total life-cycle costs and some £7 billion in total monetized benefits. This is far lower than the benefits for all other departments' projects because of the inherent limitations on quantifying—let alone monetizing—such intangibles as deterrence or security.

The MoD, like other departments, is required to submit an integrated assurance and approval plan for each project. The IPA then collects data and conducts reviews on the project throughout its life cycle.[113] For transparency, the IPA publishes annual reports on the GMPP, along with a breakdown of data on performance and risks at the project or departmental level.[114] The IPA also maintains an integrated assurance toolkit for public servants to use.[115]

Analysis of the UK's Defense Budgeting Process

In this section, we discuss the strengths and challenges of the MoD's PB&F processes, with a particular focus on possible implications for DoD.

---

[110] UK Major Projects Authority, homepage, undated.

[111] IPA, *Infrastructure and Projects Authority Mandate*, January 2021.

[112] UK Major Projects Authority, undated.

[113] IPA, 2021.

[114] IPA, *Annual Report on Major Projects, 2021–22*, July 2022; UK Infrastructure and Projects Authority and UK Cabinet Office, "Major Projects Data," webpage, updated July 20, 2023.

[115] UK Infrastructure and Projects Authority and UK Cabinet Office, "Infrastructure and Projects Authority: Assurance Review Toolkit," webpage, updated July 15, 2021.

Strengths

The MoD's PB&F processes have six strengths:

- using resources thoughtfully and responsibly
- prioritizing warfighter and mission needs
- linking plans to budgets at multiple levels
- sustaining funding for long-term initiatives
- offering flexibility for emerging requirements
- balancing oversight and agility.

Using Resources Thoughtfully and Responsibly

The MoD's Financial and Military Capability team and others in the MoD Head Office draw on a variety of data sources and analyses to inform decisions about how to allocate finite resources across the TLBs, including building on the TLB owners' proposals and engaging in negotiations and appraisals. Commonly, this process involves a cost-effectiveness analysis and a combined operational effectiveness and investment appraisal, given the difficulty in monetizing many defense outputs and benefits.

In addition, the NAO and House of Commons Public Accounts Committee annually audit the MoD and its programs, with additional layers of cost management and assurance provided by internal MoD or DE&S functions (e.g., CAAS, SSRO) and external specialist bodies (e.g., the IPA). The NAO does not engage in the political aspects of the budget but simply assesses whether the money allocated can reasonably support the strategy defined by the government in the Integrated Review and Command Paper.[116]

Through such measures, it is expected that the MoD will spend thoughtfully, responsibly, and with reduced partisan political interference. MPs appear to be less concerned than their U.S. counterparts in Congress about where defense production occurs, perhaps because construction sites are well-established and production does not influence MoD decisions as substantively.[117]

Prioritizing Warfighter and Mission Needs

The MoD system seeks to add value to the warfighter through a combination of the following factors:

- medium-term budget predictability (owing to the three- to five-year comprehensive spending reviews)
- FLCs (the military end users) responsible for proposing the optimal balance of personnel, equipment, infrastructure, and other spending to meet their requirements
- an emphasis on integration and jointness, as promoted by

---

[116] UK subject-matter expert, interview with the authors, November 2022.

[117] UK subject-matter expert, interview with the authors, November 2022.

- the central analysis and prioritization functions of the MoD's Financial and Military Capability team and others in the MoD
- the creation of UK Strategic Command as an integrator
- measures to make military careers, education, and culture more joint
- top-down allocation and monitoring of major projects—including external assessments by the NAO, Public Accounts Committee, and IPA—to ensure that money is not being misspent.

Provided the government strategy is crafted in accordance with mission needs up front, the PB&F system benefits from a series of checks to keep expenditures aligned.

Linking Plans to Budgets at Multiple Levels

The PB&F system permits the service components (the FLCs) and related organizations (e.g., the Defence Nuclear Organisation) to link their plans to their respective TLBs assigned by the MoD.

More broadly, the Command Papers and Defence Equipment Plans are derived from the defense strategic review and link MoD plans to budgets. In addition, there are lower-level processes in place to translate these high-level documents into the MoD's annual Single Departmental Plan, any subordinate strategies as needed (by DLOD or function), and the command plans and corporate plans that are developed, implemented, and reviewed at the TLB level.

Sustaining Funding for Long-Term Initiatives

MoD system programs are normally guaranteed funding at specific levels for three to five years through the defense strategic review, Defence Equipment Plan, and other strategic documents, with estimates out to ten years. The nature of the UK's parliamentary system means that there is little or no risk of parliament interfering in the specifics of the MoD's budget or delaying approvals; in any case, the automatic preauthorization of a portion of spending based on the previous year's approved expenditure allows the MoD to avoid a U.S.-style situation of political deadlock, budget sequestration, or continuing resolutions.

In contrast to the MoD's three- to five-year funding guarantees, Congress must revisit and vote on DoD's entire budget, including ongoing operational and sustainment costs, every year (albeit requiring a five-year defense plan). Certain U.S. contracts, including for munitions and missiles, must also be renegotiated every year, something that the MoD's three- to five-year stability prevents.

In recent years, the UK has also aimed to give industry greater transparency on long-term spending levels and priorities beyond the information in the Defence Equipment Plan, as well as to adopt a more collaborative approach with key industry players. Examples include the alliance construct established between the MoD and major suppliers for the *Queen Elizabeth*–class aircraft carriers and the *Dreadnought*-class nuclear-powered ballistic missile submarine programs. The UK also has a ten-year portfolio management agreement and Team Complex Weapons construct with missile manufacturer MBDA. These types of arrangements strengthen collaboration and support joint long-term planning.

Offering Flexibility for Emerging Requirements

At one level, the MoD's funding system is relatively rigid, with three- to five-year strategic defense reviews and pressure to spend money each year or risk losing it. These attributes offer a high degree of budget security but not necessarily flexibility.

At the same time, there are mechanisms for moving money between certain accounts (e.g., by virement) or for requesting additional current-year funds to meet emerging requirements.[118] Such tools have been used most recently to provide materiel and training support to Ukraine.[119]

The MoD has also, at times, proved willing to cut or cancel programs. Requirements, too, can be changed during competition.[120] It is more difficult to cancel a program once it is under contract, although it has occurred on occasion (e.g., the Nimrod maritime patrol aircraft).[121] A current example at the time of this research was the debate over whether to cancel or continue the Ajax armored vehicle project after difficulties and delays.[122] Although it is not always clear whether such cuts have been prudent from a military perspective, they do demonstrate an ability to make tough and potentially controversial changes in an evolving fiscal context or to respond to emerging threats.

Meanwhile, the MoD continues to experiment with ways of accelerating procurement timelines, including through novel contracting methods for new equipment and through the creation of various accelerators, incubators, and the MoD chief scientific adviser's Innovation Fund. The purpose of this fund is to enable flexible spending on high-risk, high-reward technologies that may fall outside traditional R&D programs.

Balancing Oversight and Agility

Each year, the MoD is externally vetted by the House of Commons Public Accounts Committee, NAO, and the Comptroller and Auditor General to ensure that funds are not misused. Audits focus on what the NAO terms the three Es: economy, efficiency, and effectiveness.[123] Within the MoD, evaluation teams undertake internal reviews of individual programs to determine risks and identify other relevant issues. Throughout the year, decisionmakers are encouraged to consider value for money and "the effective use of resources."[124] Nevertheless,

---

[118] For example, by submitting substantive SEs to parliament to increase the MoD's budget or by accessing Treasury, Deployed Military Activity Pool, or cross-governmental UKISF funds to address operational needs or operations or rapid-turnaround UCRs (MoD, 2022a, p. 22; UK subject-matter expert, interview with the authors, November 2022).

[119] UK subject-matter expert, interview with the authors, October 2022.

[120] UK subject-matter expert, interview with the authors, November 2022.

[121] UK subject-matter expert, interview with the authors, November 2022.

[122] UK subject-matter experts, interviews with the authors, November 2022.

[123] MoD, 2019b, p. 6.

[124] MoD, 2019b, p. 11.

cost overruns do occur; although they are usually handled in SEs and excess votes, these overruns can be embarrassing for the government.[125]

The MoD, like the UK government more generally, recognizes the need to scale oversight, assurance, and compliance activities to program size and risk to minimize unnecessary bureaucracy and delays. Therefore, there are additional layers of oversight for single-source contracts (via the SSRO) and major projects (via the IPA). The UK civil service also invests in developing knowledge, expertise, and thought leadership on relevant topics, such as cost modeling, risk management, P3M (project, program, and portfolio management), and assurance. Examples include the work of CAAS, the IPA's integrated assurance toolkit, the Treasury's *Green Book* and related guidance and tools, and the active participation of UK civil servants in academic and practitioner gatherings, such as the SCAF symposia.[126]

In these ways, the UK seeks to cultivate a robust but nuanced approach to oversight and assurance, balancing the risk of the misuse of funds or program difficulties and delays (because of insufficient oversight) with the risk of failing to deliver required capabilities to the warfighter in a timely manner (because of excessive caution and focus on compliance).

Challenges

Many of the challenges of the MoD's PB&F system are the flip side of its strengths, given that the UK approach represents a continuously evolving compromise between countervailing imperatives: stability versus agility, control versus flexibility, oversight versus efficiency of implementation, quality versus affordability, and so on. In sum, our review of the MoD's PB&F processes revealed three primary challenges:

- ongoing struggles to plug the MoD's budget "black hole" of cost overruns
- enduring barriers to fungibility and flexibility
- enduring barriers to rapid acquisition and innovation.

Ongoing Struggles to Plug the MoD's Budget "Black Hole" of Cost Overruns

The MoD has made substantial strides in reforming and improving its PB&F processes over the past decade. Examples include the Levene Reforms of 2011, which empowered FLCs and delivery organizations and rationalized senior posts and decisionmaking within the MoD, as well as the creation of the SSRO in 2014.

Nonetheless, there remains a persistent concern from parliament, the wider UK public, and key allies and partners (including the United States) that the MoD's budgets and plans are simply insufficient to address the threats the country faces or achieve the UK's global ambitions. Political leaders have acknowledged these issues and issued repeated calls to increase

---

[125] UK subject-matter expert, interview with the authors, November 2022.

[126] SCAF, formerly known as the Society for Cost Analysis and Forecasting, was established in the UK in 1984. See SCAF, "About Us," webpage, undated.

defense spending to 3 percent of GDP—calls that intensified in the wake of Russia's invasion of Ukraine in February 2022. However, the 2023 refresh of the Integrated Review pledges to spend only 2.25 percent by 2025, and it is unclear whether or when its aspiration to reach 2.5 percent of GDP at an unspecified later date can be made a reality given the fiscal climate.[127]

Furthermore, because the MoD budget is expressed in nominal terms, it is greatly affected by inflation in the form of both rising fuel and other running costs and the valuation of the pound sterling relative to other currencies (especially the U.S. dollar), given that the MoD often purchases assets from abroad.[128] Consequently, the MoD budget has been hit especially hard by recent inflationary pressures, exchange rate volatility, and the financial costs of Brexit.

This type of overheated MoD budget is not a recent phenomenon. It has been a topic of considerable controversy since before the financial crisis of 2008–2009; the NAO and the Defence Select Committee and Public Accounts Committee in the House of Commons have repeatedly published reports criticizing the MoD's budget and Defence Equipment Plan as equivalent to a multibillion-pound black hole.[129] Some argue that the estimates published in the Defence Equipment Plan are "badly underestimated."[130] Others suggest that the bottom-up nature of the demand signals within MoD can generate overoptimism among individual stakeholders with vested interests in keeping their programs alive, resulting in unrealistically low budgets.[131] Budgets also tend to be built without large contingencies, another structurally overoptimistic behavior in the MoD and the government more broadly, and this tendency can result in sizable overruns.[132]

Building on these themes, the Public Accounts Committee issued a blunt and damning 2021 report, concluding,

> The Ministry of Defence has once again published a ten-year military equipment and capabilities Plan with a funding "black hole" at its centre, potentially as big as £17.4 billion. . . .
>
> The MoD remains stuck in a cycle of focusing on short-term financial pressures. It has sought to balance its annual budget by again deferring or descoping the development of capabilities, resulting in poor long-term value for money and the use of all its contingency funds in 2020–21 to help offset funding shortfalls. . . .

---

[127] UK Cabinet Office, 2023.

[128] UK subject-matter expert, interview with the authors, October 2022.

[129] Helen Warrell, "MoD Accused of Overspending as Budget 'Black Hole' Hits £17bn," *Financial Times*, January 12, 2021.

[130] UK subject-matter expert, interview with the authors, November 2022.

[131] UK subject-matter expert, interview with the authors, November 2022.

[132] UK subject-matter expert, interview with the authors, November 2022.

Once it has finally established a balanced equipment programme, MoD will "need to develop a more sophisticated approach to assessing future funding pressures and managing its equipment expenditure."[133]

For its part, the NAO's subsequent 2022 report on the more-recent annual iteration of the Defence Equipment Plan, covering the period 2022–2032, stated that the revised plan was affordable and reflected "ongoing improvements" by the MoD. However, the NAO noted that the plan was "based on optimistic assumptions that it will achieve all planned savings" and that, given recent events in Ukraine and the global economy, "the volatile external environment means this year's Plan is already out of date."[134] Therefore, the NAO emphasized that

> The Department faces significant and growing cost pressures which will have an immediate impact on its spending plans. The Department believes it can manage these pressures but has left itself limited flexibility to absorb any cost increases on equipment projects, or across other budgets. It needs to address the financial challenges promptly to avoid falling back into old habits of short-term cost management, which do not support longer-term value for money. The cost pressures are also likely to undermine the pace at which it can modernize the Armed Forces. The Department will need to make difficult prioritization decisions to live within its means and retain enough flexibility in its Plan to respond promptly to changing threats.[135]

The MoD has taken steps over the past decade to address the gap between its ambitions and resources, including efficiency savings by cuts to personnel in the MoD Head Office, cuts to capability and troop numbers, an increased emphasis on partnering and off-the-shelf products (where possible) to drive down costs, and reforms to PB&F processes themselves. However, these steps have arguably been insufficient to reconcile the UK's global ambitions with its finite resources as a medium power. Therefore, a resolution to this enduring challenge may require a broader shift in the UK's public debate about its international role, especially post-Brexit, and in its appetite to scale its defense spending accordingly—rather than merely continuing the cycle of efficiencies and reforms at the MoD level.[136]

Enduring Barriers to Fungibility and Flexibility

Although the procedure through which parliament formally funds the MoD—three- to five-year comprehensive spending reviews and correlated strategic defense reviews—is a strength,

---

[133] UK House of Commons, Public Accounts Committee, "New Defence Money Potentially Lost in 'Funding Black Hole' at Centre of UK Defence Equipment Plan," March 16, 2021.

[134] NAO, *The Equipment Plan, 2022 to 2032: Ministry of Defence*, November 29, 2022b, p. 9.

[135] NAO, 2022b, p. 10.

[136] Hew Strachan and Ruth Harris, *The Utility of Military Force and Public Understanding in Today's Britain*, RAND Corporation, RR-A213-1, 2020.

it is also a potential Achilles' heel. On one hand, it provides a high degree of certainty, stability, and transparency to the UK government and public. On the other hand, it is relatively rigid.

In theory, there are mechanisms for moving money between commodity blocks and even years; in practice, however, a use-it-or-lose-it culture still applies to most funding and drives inefficiencies in annual spending. There is no incentive to underspend funds but instead a perverse incentive to spend everything. Underspending funds can lead to at least two negative outcomes: a race to spend unused funds near the end of the fiscal year, which leads to inefficiencies, or a budgetary reduction during the next spending review (a penalty).[137]

Enduring Barriers to Rapid Acquisition and Innovation

Although the MoD can draw on mechanisms to hasten acquisition and promote innovation, such as UCRs, and although it has launched multiple reform initiatives in recent years, there are still persistent criticisms of the speed of the acquisition cycle, criticisms of the failure to adequately fund R&D and promote exports, and barriers to rapid delivery, experimentation, and innovation.[138] Making the MoD more agile and overcoming associated structural, process, funding, skill alignment, and cultural challenges remains a work in progress. It certainly cannot be said that the MoD has solved all these problems, which are common to DoD and other defense establishments.[139] The need to accelerate acquisition and innovation is a known and urgent challenge for the MoD that it continues to address in light of the threats posed by Russia and China. In an era of strategic competition, the MoD must prepare for an increased tempo and variety of operations.

Applicability

More than the other U.S. allies considered in this study (which have smaller defense budgets and fewer missions), the UK faces many challenges that are similar to those of the United States. Despite an order-of-magnitude difference in budget, there are parallels between the MoD and DoD in terms of the need to fund, develop, and field full-spectrum capabilities and a joint force optimized to conduct expeditionary operations around the globe.

Much of the MoD's experience could be applicable to DoD—whether as a useful comparator, an inspiration for possible changes, or a cautionary tale about the pitfalls of certain models. The most obvious exceptions are aspects of the UK's approach to PPBE (i.e., PB&F) that are not relevant to DoD because of the sizable differences between the UK and U.S. constitutions, especially the relationship between the executive and legislative branches.

Given the UK's status as one of the closest U.S. allies and the fact that the MoD and DoD face many of the same challenges, there is a clear benefit in continuing to share transatlantic

---

[137] UK subject-matter expert, interview with the authors, October 2022.

[138] Rebecca Lucas, Lucia Retter, and Benedict Wilkinson, *Realising the Promise of the Defence and Security Industrial Strategy in R&D and Exports*, RAND Corporation, PE-A2392-1, November 2022.

[139] Retter et al., 2021.

perspectives on PPBE and PB&F. This sharing could entail ongoing exchanges in which DoD officials work closely with MoD counterparts to share experiences and best practices as both countries embark on continuing reforms to their defense budgeting processes.[140]

Lessons from the UK's Defense Budgeting Process

The UK's defense budgeting process likely holds useful insights for DoD in several areas.

Lesson 1: The UK's Defense Budgeting Process Strives for a Virtuous Cycle with Industry

With the publication of each year's Defence Equipment Plan, the MoD estimates its ten-year procurement priorities, sending annual demand signals to guide long-term industrial investment, training, and R&D. The MoD also works closely with a small group of large defense companies that can deliver highly complex systems while involving smaller enterprises in their supply chains. The MoD aims for a virtuous cycle whereby its R&D or procurement funding can help the private sector develop new technologies that can then be used by the MoD and be exported abroad for the overall benefit of the UK.

Lesson 2: The UK's Defense Budgeting Process Guarantees Long-Term Programming

MoD programs are normally guaranteed funding at specific levels for three to five years through the defense strategic review, Defence Equipment Plan, and other strategic documents, with estimates out to ten years. The automatic preauthorization of a portion of spending based on the previous year's approved expenditure allows the MoD to avoid a U.S.-style situation of political deadlock, budget sequestration, or continuing resolutions. In contrast to the MoD's multiyear funding guarantees, Congress must revisit and vote on DoD's entire budget every year. Certain U.S. contracts, including for munitions and missiles, must also be renegotiated every year, which the MoD's programming stability prevents. Multiyear spending reviews make budgeting more rigid than a yearly budget would, but HM Treasury and MoD retain some flexibility when translating medium-term spending plans into annual budgets.

---

[140] There might be utility in similar exchanges between, for example, Congress and parliament or the U.S. Government Accountability Office and the NAO.

Lesson 3: The UK's Defense Budgeting Process Attempts to Balance Decentralization with Jointness and Multidomain Integration, While Also Encouraging Innovation

The UK's PB&F system benefits from cross-governmental mechanisms and joint funds, such as the UKISF, that allow it to allocate resources to urgent requirements in a contingency while incentivizing interagency work. Such mechanisms and funding sources allow the MoD to address the root causes of conflict and instability rather than merely reacting to them militarily. These efforts demonstrate ways to balance decentralization with organizational, process, and cultural measures that promote jointness and multidomain integration. They also demonstrate how broader changes to institutional and individual culture can combat the effects of interservice rivalries. The UK's PB&F system also benefits from novel funding structures and processes for encouraging innovation and rapid acquisition. Notable examples of these novel structures and funding arrangements include the Innovation Fund and the Defence and Security Accelerator.

Lesson 4: UK and U.S. Defense Budget Reform Efforts Could Inform One Another

The UK's expeditionary focus, global commitments (like those of the United States), and requirements for diverse capabilities for diverse terrains set it apart from other medium powers. The mixture of long-term and immediate pressures poses significant dilemmas for UK defense planners and defense budget managers. But the pressures also provide an impetus for ongoing efforts to adapt the MoD's PB&F processes to encourage more agility and innovation, improve value for money, and enable the MoD to deliver increased output despite limited resources. Given the UK's status as one of the closest U.S. allies and the fact that the MoD and DoD face many of the same challenges, there would be a clear benefit in sharing transatlantic perspectives on PPBE and PB&F, perhaps in the form of ongoing exchanges between DoD officials and their MoD counterparts to share experiences and best practices as both countries embark on continuing reforms to their defense budgeting processes. Similar useful exchanges could occur between Congress and parliament or between the U.S. Government Accountability Office and the NAO.

Table 4.1 summarizes these lessons.

TABLE 4.1

Lessons from the UK's Defense Budgeting Process

Theme	Lesson Learned	Description
Decisionmakers and stakeholders	Lesson 1: The UK's defense budgeting process strives for a virtuous cycle with industry.	In its ten-year procurement plans, the MoD sends annual demand signals to guide long-term industrial investment, training, and R&D in new technologies that can benefit the MoD and the UK.
Planning and programming	Lesson 2: The UK's defense budgeting process guarantees long-term programming.	MoD programs are usually guaranteed funding for three to five years, with estimates out to ten years, plus automatic preauthorization of a portion of the previous year's approved expenditure.
Budgeting and execution	Lesson 3: The UK's defense budgeting process attempts to balance decentralization with jointness and multidomain integration, while also encouraging innovation.	Cross-governmental mechanisms and joint funds allow the UK system to allocate resources to urgent requirements while incentivizing interagency work. The UK also benefits from novel funding processes for encouraging innovation and rapid acquisition.
Oversight	Lesson 4: UK and U.S. defense budget reform efforts could inform one another.	Given that the UK is one of the closest U.S. allies and that the MoD and DoD face many of the same challenges, there would be a clear benefit in sharing transatlantic perspectives on PPBE and PB&F reforms at various levels of government.

CHAPTER 5

Key Insights from Case Studies of Selected Allied and Partner Nations

The 2022 NDS focuses on strengthening relationships with allied and partner nations. In this chapter, we distill our insights into how the defense resource planning processes of selected allies—Australia, Canada, and the UK—support the pursuit of shared strategic goals. There is a clear benefit for the United States and its allied and partner nations in sharing their perspectives on defense resource planning processes, particularly their ways of balancing speed and oversight while promoting innovation.

In Volume 1, we discussed how China and Russia conduct defense resource planning, programming, budgeting, execution, and oversight—and the strengths and challenges of their approaches, albeit with imperfect information. For this volume, we had ample material available to describe in detail the processes of Australia, Canada, and the UK in Chapters 2 through 4.

This final chapter focuses on summary takeaways. As part of this analysis, we used an initial set of standard questions from the Commission on PPBE Reform, focusing on core areas related to resource planning, to ensure that there would be some ability to compare across cases. The material presented in this chapter, which is drawn from Chapters 2 through 4, distills important themes for the commission to understand when trying to compare the U.S. defense resource planning process with those of selected allies and partner nations. The similarities of the Australian, Canadian, and British processes to those of the U.S. PPBE process far exceed the differences and therefore suggest multiple insights that are germane for the United States.

The following section on key insights consolidates the strengths, challenges, and lessons outlined in the case studies in this volume. The concluding section on applicability speaks directly to the commission's mandate—and to the potential utility of these insights for DoD's PPBE System.

Key Insights

Australia, Canada, and the United Kingdom Have a Shared Commitment to Democratic Institutions with the United States and Converge on a Similar Strategic Vision

The United States, Australia, Canada, and the UK have similar strategic visions and aim to counter similar strategic threats. This alignment presents opportunities for co-development and broader opportunities to work together toward shared goals, but also it requires the United States and its allies to develop plans and processes to facilitate more-effective partnership approaches. In addition, each country struggles to balance at least four often-competing priorities: keeping pace with strategic threats, executing longer-term plans, using deliberate processes with sufficient oversight, and encouraging innovation.

Australia's Defence operates in close concert with several allies, especially the United States, and leverages those alliances and partnerships as a central tool of national security. Australia is a member of the Five Eyes (Australia, Canada, New Zealand, UK, and United States) security agreement, the Quadrilateral Security Dialogue ("the Quad"), and the AUKUS agreement. Australia is a strategically located partner in the Indo-Pacific theater and shares U.S. concerns about China's military rise.

Canada and the United States have a long, collaborative defense relationship. Their militaries have fought alongside one another in several conflicts since World War II. Both countries are members of NATO, and they cooperate extensively through several bilateral defense forums and agreements, including NORAD, the Permanent Joint Board on Defense, the Military Cooperation Committee, the Combined Defense Plan, the Tri-Command Framework, the Canada-U.S. Civil Assistance Plan, and the National Technology Industrial Base.[1] Canada recognizes its relatively small military size on the world stage and emphasizes cooperation with allies; specifically, the United States and NATO member countries. However, Canada lacks the population and military personnel to sustain large overseas military deployments, and, thus, its 2017 policy limits the size and duration of planned contributions. Nonetheless, CAF participates throughout the year—again, largely with allies—in operations and joint military exercises, including assurance missions, stability operations, and United Nations missions.[2]

The UK is a critical U.S. ally that retains global military responsibilities and capabilities, including nuclear weapons. The UK is a member of the AUKUS security pact, the Combined Joint Expeditionary Force with France, the European Intervention Initiative, the Five Eyes security agreement, the Five Power Defence Arrangements (with Australia, Malaysia, New Zealand, and Singapore), the Joint Expeditionary Force (which it leads), NATO, and the Northern Group. It is also a veto-wielding permanent member of the United Nations Secu-

---

[1] Government of Canada, 2014.

[2] For a list, see DND, 2022b.

rity Council. Therefore, the MoD interacts frequently and interoperates closely with the U.S. military and intelligence community, and its defense budget and planning decisions are often made in unofficial concert with DoD decisions and priorities.

Foreign Military Sales Are an Important Mechanism for Strategic Convergence but Pose Myriad Challenges for Coordination and Resource Planning

Australia, Canada, and the UK rely on U.S. FMS to promote strategic convergence, interconnectedness, interoperability, and interchangeability. One downside to this reliance is exchange rate volatility, which can require budget adjustments to cover exchange rate adjustments. Accrual-based accounting for life-cycle planning can present additional challenges when adjusting FMS for inflation or exchange rates; these challenges can also require revisions to cost estimates. However, given the strong allied focus of the defense strategies of all these countries, each country places a heavy emphasis on the importance of interoperability and, in some cases, integration—a key consideration in acquisition and force generation. This strategic emphasis poses a further challenge to each country's ability to independently act with flexibility.

Australia's Defence is typically a customer of U.S. systems, often through FMS. Because the development and production of these systems may depend on DoD's PPBE processes, there are limitations to Australia's ability to become more agile than those U.S. processes will allow—at least with respect to major weapon systems. This constraint is acceptable to Defence in view of the interoperability and capability advantages. In discussions about AUKUS, emerging technologies, innovation, and weapon cooperation, our interviewees indicated that the U.S.-Australia relationship may shift to one in which Australia is not simply a defense materiel customer but more of a partner. Beyond technology cooperation, there is the prospect that greater transparency and coordination across U.S. and Australian PPBE processes could lead to mutual benefits in terms of capability agility, synergies, and efficiencies.

Canada also relies on U.S. FMS. Major contracting and FMS for DND (and other Canadian federal agencies) are handled by Public Services and Procurement Canada, which centralizes major purchases for the Canadian government. This centralization may cause internal delays in processing FMS. Along with the exchange rate volatility that Australia and the UK experience, the Canadian fiscal year does not align with the U.S. fiscal year, which can cause additional problems when planning or accounting for FMS purchases.

In the UK, inflationary pressures have been exacerbated by a strengthening U.S. dollar, to which the MoD is especially sensitive because of its large number of major U.S. FMS contracts and some fixed-price fuel-swap contracts denominated in dollars.[3] The UK defense sector is highly exposed to exchange rate volatility, given the extent of its U.S. imports, primarily air-

---

[3]   The MoD maintains multiple euro and dollar bank accounts and enters into forward-purchase contracts for these currencies to mitigate the risk from changing exchange rates.

craft (e.g., F-35Bs, P-8s, AH-64 *Apache* helicopters, CH-47 *Chinook* helicopters). Because the MoD budget is expressed in nominal terms and because the MoD often purchases assets from abroad, it is greatly affected by inflation in fuel and other running costs and by the valuation of the pound sterling relative to other currencies (especially the U.S. dollar).[4] Consequently, the MoD budget has been hit especially hard by recent inflationary pressures, exchange rate volatility, and the financial costs of Brexit.

The Australian, Canadian, and UK Political Systems Shape the Roles and Contours of Resource Planning

The political systems in the three allied countries are similar in that the executive branch has the power of the purse, which reduces political friction over appropriations.

The Australian electorate votes for parties, not prime ministers. A prime minister is selected by the party that holds the majority in the new government, and, subsequently, the prime minister appoints senior elected colleagues to ministerial positions, which are comparable to secretaries in the U.S. cabinet. Each minister is therefore an elected official. Under the Minister for Defence, there is both a departmental secretary—who is a career bureaucrat rather than a political appointee—and the CDF, who is a military officer.[5] Overall, the legislative and executive branches of the Australian system are more closely linked than they are in the United States.

The two major parties in Australia take a relatively bipartisan approach to defense; hence, a change of government does not necessarily result in any significant change in defense plans or budget allocations. New governments sometimes direct the department to begin work on a new defense white paper; however, such changes in strategic guidance are typically related more to changes in the geostrategic environment than to politics.[6]

In Canada, members of the lower chamber of parliament, the House of Commons, are elected by voters; the leader of the largest party in that chamber becomes prime minister and selects the cabinet. The executive branch plays the dominant role in Canada's budget preparation, and parliament has relatively limited influence. Parliament performs legislative and oversight functions through its review and approval of the budget.[7] When the executive controls a majority of seats in the House of Commons, it is in a very strong position to have its prepared budget approved with minimal or no changes. When the executive controls a plurality of seats but not a majority, it relies on support from the opposition or other, smaller

---

[4] UK subject-matter expert, interview with the authors, October 2022.

[5] The CDF is Australia's senior military officer, the only four-star officer in the ADF. The CDF leads the integrated Australian Department of Defence and ADF as a diarchy with the Defence Secretary.

[6] The 2022 change in government leadership from the Liberal to the Labour party did not result in any substantial change in priorities or budget for the Australian Department of Defence, nor did it result in a new defense white paper.

[7] Armit, 2005, p. 2.

parties to pass budgets and other key legislation. If the ruling government cannot pass its budget through the House of Commons, it is considered to no longer have the confidence of the House, and an election must be called.

In the UK, the stability of the bicameral parliamentary system relies on the fact that the chief of the executive branch (the prime minister) is an elected member of parliament from the party with a majority in the elected lower chamber (the House of Commons). Members of the upper chamber, the House of Lords, are not elected but appointed. By centuries-old convention, the upper chamber defers to the lower chamber on financial matters. Because the UK does not have a codified constitution and instead relies to a significant extent on accumulated convention, there is less inherent antagonism between the branches of government than in the United States. The resulting empowerment of the prime minister can enable more-streamlined executive and legislative action, but it also limits the formal checks and balances that characterize the U.S. system.

Parliament must approve the government defense missions and the resources that the MoD requests for those missions. Opposition from the prime minister's own majority in the House of Commons triggers a no-confidence vote and the likely collapse of the government. The alignment of resource allocation with the MoD's mission is therefore a structural feature of the UK parliamentary system.

Australia, Canada, and the UK Have Less Legislative Intervention in Budgeting Processes, Relative to the United States, and No Continuing Resolutions

The Australian, Canadian, and UK resource management systems have less partisan interference than in the United States, according to subject-matter experts on these systems.

Compared with DoD, Australia's Department of Defence receives significantly less PPBE guidance from the Australian legislature. The executive branch—the Minister for Defence, the prime minister, and cabinet colleagues—hold the purse strings. The other MPs and senators can review Australia's PPBE-like functions and direct their questions to either the Minister for Defence or directly to the Australia's Department of Defence through parliamentary liaison officers. Unlike in the United States, the annual budget for existing services and programs in Australia appears in a separate appropriations bill from that for new programs,[8] making it unlikely that existing government services will be blocked and effectively eliminating any need for a continuing resolution.

Canada's government is never at risk of a shutdown because of funding lapses. Parliament can enact interim estimates that authorize spending at proposed levels until the Main Estimates pass through the normal legislative process, or the executive can take other extraordinary measures to continue funding ongoing government functions.

---

[8] For details on the separation of appropriations bills for continuing services and new policies, see Webster, 2014.

The UK parliamentary system offers little or no risk of parliament interfering in the specifics of the MoD's budget or delaying approvals. In any case, the automatic preauthorization of a portion of defense spending based on the previous year's approved expenditure allows the MoD to avoid a U.S.-style situation of political deadlock, budget sequestration, or continuing resolutions. MPs appear to be less concerned than their U.S. counterparts in Congress with where defense production occurs, perhaps because construction sites are well-established and production does not substantively influence MoD decisions.[9]

Strategic Planning Mechanisms in Australia, Canada, and the UK Harness Defense Spending Priorities and Drive Budget Execution

Australia, Canada, and the UK each start their defense resource management processes with strategic planning that tries to identify key priorities for finite funds in defense budgets that are smaller than that of the United States.

Australia's defense budgeting system is guided by a series of strategic planning documents, such as the *2016 Defence White Paper* and *2020 Defence Strategic Update*,[10] that lay out strategic goals, capability priorities, and funding profiles for the following decade. The Portfolio Budget Statement and the IIP, both of which are derived from mission needs and strategic priorities, reflect the "value to the warfighter" of resource allocations. The DCAP ensures that the current and planned force structure is fit for prospective operational scenarios, theater campaign plans, operational concepts, and preparedness directives.

Canada's defense programs are also based on several strategic planning documents: 2017's *Strong, Secure, Engaged*; 2018's *Defence Investment Plan* and *Defence Plan, 2018–2023*; 2019's updated *Defence Investment Plan*; 2020's *Defence Capabilities Blueprint*; and the latest *Department of National Defence and Canadian Armed Forces Departmental Plan*, which was released in 2022.[11] Together, these strategic documents provide the basis for defense budgeting decisions.

The MoD's approach to strategic planning begins with its mission as outlined in the Defence Command Paper. The most recent iteration of this white paper, *Defence in a Competitive Age*, was published in March 2021.[12] The white paper states that the seven primary goals of the MoD and the British Armed Forces are to defend the UK and its overseas territories, sustain the country's nuclear deterrence capacity, project the UK's global influence, execute its NATO responsibilities,[13] promote national prosperity, contribute to peacekeeping, and support the defense and intelligence-gathering capabilities of the UK's allies and part-

---

[9] UK subject-matter expert, interview with the authors, November 2022.

[10] Australian Department of Defence, 2016b; Australian Department of Defence, 2020a; Brangwin and Watt, 2022.

[11] DND, 2017; DND, 2018a; DND, 2018b; DND, 2019; DND, 2020a; DND, 2022a.

[12] MoD, 2021a; UK Cabinet Office, 2021.

[13] UK House of Commons, Defence Select Committee, 2021, p. 1.

ners.[14] The Treasury aligns fiscal resources to support these missions through comprehensive spending reviews.

Jointness in Resource Planning Appears to Be Easier in Australia, Canada, and the UK Given the Smaller Size and Structure of Their Militaries

In Australia, the ADF operates in a relatively more joint manner than its U.S. counterpart. Some ADF program costs, such as fuel costs, are centralized.[15] There is a level of joint financial governance; service component CFOs report to the departmental CFO and to their service chiefs. These points may be important to the U.S. defense community, given ongoing efforts to enhance jointness across the U.S. military.

In Canada, military service acquisition projects are managed by a DND process that is service-agnostic and ranks projects according to DND priorities. This process ensures that service-centric views do not dominate procurement planning and encourages more collaboration, discussion, and consensus.

With cross-governmental mechanisms and joint funds, such as the UKISF, the UK's PB&F system can allocate resources to urgent requirements while incentivizing interagency work. Such mechanisms and funding sources allow the MoD to address the root causes of conflict and instability rather than merely reacting to them militarily. These efforts demonstrate ways to balance decentralization with organizational, process, and cultural measures that promote jointness and multidomain integration. They also demonstrate how broader changes to institutional and individual culture can combat the effects of interservice rivalries.

Australia, Canada, and the UK Place a Greater Emphasis on Budget Predictability and Stability Than on Agility

Australia's Defence is given assurance of sustained funding levels over a four-year rolling period. The *2016 Defence White Paper* laid out a baseline for defense spending over ten years. The *2020 Defence Strategic Update* laid out an updated version of this baseline, extending it to 2029–2030. Defence plans its investments out as far as 20 years as whole-of-life investments.[16]

Canada's notional DND budget is guaranteed to continue year on year, allowing for better decisionmaking in out-years. DND's Capital Investment Fund ensures that approved projects will be paid for years or even decades to come, regardless of a change in government.

MoD programs are normally guaranteed funding for three to five years, with estimates out to ten years. In contrast, with only a few exceptions, Congress must revisit and vote on DoD's entire budget every year (albeit requiring a five-year defense plan). Certain U.S. con-

---

[14] MoD, undated-a.

[15] Australian Defence official, interview with the authors, November 2022.

[16] Australian Defence officials, interviews with the authors, October and November 2022.

tracts, including for munitions and missiles, must also be renegotiated every year, something that the MoD avoids. These attributes of the systems of Australia, Canada, and the UK offer a high degree of budget security but not necessarily flexibility.

Despite the Common Emphasis on Stability, Each System Provides Some Budget Flexibility to Address Unanticipated Changes

The Australian Parliament can boost the defense budget in periods of national emergency (e.g., wildfires) or overseas military operations (e.g., Iraq, Timor-Leste) using the no-win/no-loss model for deployments.[17] The government can supplement Defence's allocation to alleviate inflationary pressures. The NSC can consider urgent priorities and their funding implications, and the Minister for Defence can intervene to prioritize certain programs or investments. There is flexibility to move current-year funds among groups and military services to meet emerging needs. The CFO can divert funding to meet emerging priorities. The DCAP promotes agility, but it is linked to government updates of strategic guidance, which may not be sufficiently agile.[18] However, there has been an effort to make these updates more frequent and ongoing, and there is an intent for capability processes to be more agile in the future, in cases where reducing operational risk is more important than acquisition risk.[19]

In the DND, regular supplementary parliamentary spending periods can help close unforeseen funding gaps for emerging requirements and help manage risk. DND officials believe that planning capital investments on an accrual basis while managing year-on-year funding on a cash basis allows for more-flexible funding. DND does not require parliamentary approval—nor must it inform parliament—to transfer funds within a vote from one program to another. DND can carry forward to the next fiscal year up to 5 percent of total operating expenditures, which it can use to adjust misalignments in spending.[20]

For the MoD, multiyear spending reviews make budgeting more rigid than a yearly budget would, but the Treasury and the MoD retain some flexibility when translating the spending reviews into annual budgets and plans. The UK also has mechanisms for moving money between accounts and accessing additional funds in a given fiscal year. These mechanisms include a process known as virement for reallocating funds with either Treasury or parliamentary approval. The MoD can make additional funding requests through in-year supplementary estimates sent to parliament. The MoD has access to additional Treasury funds to cover UCRs, and it can use the cross-governmental UKISF or the Deployed Military Activity

---

[17] No-win/no-loss funding is appropriated through appropriations bills. It can be appropriated to offset the cost of approved operations and foreign exchange movements.

[18] Australian Defence official, interview with the authors, October 2022.

[19] Australian Defence official, interview with the authors, October 2022.

[20] Perry, 2015.

Pool "to make available resources to fund the initial and short-term costs of unforeseen military activity,"[21] such as responses to natural disasters or support to Ukraine.[22]

Similar Budget Mechanisms Are Used in Australia, Canada, and the UK

Australia, Canada, and the UK use similar budget mechanisms, including the carryover of funds, movement of funds across portfolios, appropriations with different expirations, and supplementary funds for emerging needs. The use of these mechanisms, however, varies across the cases.

Australia's Defence has five key cost categories, which are similar to U.S. appropriation categories: workforce, operations, capability acquisition program (including R&D), capability sustainment, and operating costs.[23] There is limited movement among categories, but there is flexibility for "unders" and "overs," meaning that funds can be shifted from categories with a surplus to categories with a deficit. Projects are funded and managed on a whole-of-life basis,[24] accounting for both capital and operating costs. Under the no-win/no-loss mechanism for deployments,[25] Defence is reimbursed for most operational costs and must return unused funds to the Treasury. Defence absorbs some level of its costs, but the majority is offset by government reimbursement.

Within Australia's Defence IIP, both approved government projects and unapproved, fungible programs can be shifted "left" or "right" (accelerated or delayed) as needs arise.[26] To manage the risk of underachievement (or overexpenditure), the IIP is 20-percent overprogrammed for acquisition in the current financial year. Other types of Australian funding are also fungible in that they can be shifted across the defense portfolio, including across groups and military services. The operating budget for Defence expires at the end of each financial year, but major procurements are handled separately through the IIP and do not expire. Still, the overall acquisition program is expected to hit a target annual expenditure level.

When the Canadian Minister of Finance presents the annual national budget to the House of Commons, there are one or more votes that correspond roughly to different colors of money. Each color of money is assigned an arbitrary, noncontiguous vote number. Common votes include vote 1 for operating expenditures, vote 5 for capital expenditures, vote 10 for grants

---

[21] MoD, 2022a, p. 22.

[22] UK Cabinet Office, 2023.

[23] In this context, *operating* relates to the forecasted costs to support defense systems, including training on those systems, whereas *operations* relates to nonforecasted costs associated with deployed forces.

[24] Under the ODCS, approval to acquire new weapon systems requires an estimate of total costs through the system's projected end of life, including personnel, operating, and sustainment costs.

[25] This is not to be confused with the day-to-day running of the ADF.

[26] Australian Defence official, interview with the authors, October 2022; Australian Department of Defence, 2020c.

and contributions, and vote 15 for long-term disability and life insurance plans. The votes can span a portfolio of programs or apply to specific programs. Organizations can transfer funds within a vote from one program to another without parliament's approval.[27] Organizations *do* need parliament's approval to transfer funds *between* votes (e.g., from vote 1 to vote 5). Canadian federal agencies can also carry forward a portion of their unspent funds from a previous year, typically up to 5 percent of operating expenditures and 20 percent of capital expenditures.[28]

The UK uses both accrual-based budgeting (based on when transactions occur rather than when cash receipts or payments are exchanged) and zero-based budgeting (in which all activities and programs must be recosted from zero and justified through a set of criteria for prioritizing projects with the highest value for money). The Treasury controls the MoD's spending using the accrual system.[29] The MoD reports its spending monthly to comply with Treasury reporting requirements.[30] Like those of every other department, the MoD's budget works on a "spend-it-or-lose-it" basis by which the money allocated each year must be spent or it is returned without compensation.

Australia, Canada, and the UK Have All Pivoted Toward Supporting Agility and Innovation in the Face of Lengthy Acquisition Cycles

Australia's Defence has been looking for ways to increase agility. One way would extend the no-win/no-loss provision for operations to ordering ordnance and other expendables prior to a conflict so that the ADF would be more prepared for emerging threats. To accelerate innovation, the proposed ASCA would be required to secure funds for capabilities in which technologies arise faster than capability planning time frames with greater agility, efficiency, and effectiveness.

Like DoD, Australia's Department of Defence possesses technology facilitators, such as DIH, that help integrate emerging technologies with defense priorities. But there are few examples of the successful adoption of new innovations through DIH. And although the goal of ASCA is to help fast-track innovations into service, some observers acknowledge that that agency's success will be highly dependent on broader changes to PPBE-like processes to facilitate agility.

Canada's strategic plan states that the DND should exploit defense innovation as a priority.[31] Canada is working with the United States on NORAD modernization as one of its priorities.

[27] Pu and Smith, 2021.

[28] Pu and Smith, 2021.

[29] UK subject-matter experts, interviews with the authors, October and November 2022.

[30] MoD, 2019b, p. 26.

[31] DND, 2022a.

Like DoD, the MoD is experimenting with new ways to encourage innovation, including a new dedicated Innovation Fund, which allows the chief scientific adviser to pursue higher-risk projects as part of the main R&D budget. The MoD has also been experimenting with ways to accelerate procurement timelines, including through novel contracting methods for new equipment and through the creation of various accelerators and incubators.

Australia, Canada, and the UK Have Independent Oversight Functions for Ensuring the Transparency, Audits, or Contestability of Budgeting Processes

Accountability in Australia is provided through several means: the Australian National Audit Office, the Portfolio Budget Statement, the contestability function, and other reviews. The Australian National Audit Office, as an independent auditor, is similar to the U.S. Government Accountability Office and the NAO. The Portfolio Budget Statement is subject to public and parliamentary scrutiny; although the opposition can rarely change the spending decisions presented in the statement, public grievances can be aired, thereby pressuring the government as elections loom. The contestability function informs oversight but is not oversight itself; rather, contestability advice is integrated into the decisionmaking of the Defence Investment Committee, the Defence Finance and Resources Committee, and the NSC. Oversight also exists through independent reviews of acquisition activities and through Senate reviews of defense programs.

Parliamentary oversight—or scrutiny—in Canada is aided by analyses from the Auditor General, the Parliamentary Budget Officer, and, at times, the Library of Parliament. The former two roles are accountable directly to parliament rather than to the executive or a minister. The Auditor General holds office for a ten-year term, issues an annual report to the House of Commons, produces other audits during the year, and appears regularly before parliamentary committees.[32] The Parliamentary Budget Officer holds office for a seven-year term and provides estimates on matters relating to Canada's finances or economy either independently or at the request of a parliamentary committee. The Parliamentary Budget Officer issues an annual report to both chambers of parliament in addition to reports requested by committees or parliamentarians, all of which are meant to raise the quality of debate and promote budget transparency. At the beginning of each fiscal year, the Parliamentary Budget Officer also submits an annual work plan with a list of matters that the office intends to bring to the attention of parliament.[33]

Each year, the MoD is externally vetted by the House of Commons Public Accounts Committee, the NAO, and the Comptroller and Auditor General to ensure that funds are not misused. Audits focus on what the NAO terms the three Es: economy, efficiency, and effective-

---

[32] Barnes, 2021, p. 2.

[33] Barnes, 2021, pp. 9–10.

ness.[34] Within the MoD, evaluation teams undertake internal reviews of individual programs to determine risks and identify other relevant issues. Throughout the year, decisionmakers are encouraged to consider value for money and "the effective use of resources."[35] Nevertheless, cost overruns do occur, and they can be embarrassing for the government.[36]

The MoD recognizes the need to scale oversight, assurance, and compliance activities to program size and risk to minimize unnecessary bureaucracy and delays. Therefore, there are additional layers of oversight for single-source contracts and major projects. In these and other ways, the UK seeks to cultivate a robust but nuanced approach to oversight and assurance, balancing the risk of the misuse of funds or program difficulties and delays (because of insufficient oversight) against the risk of failing to deliver required capabilities to the warfighter in a timely manner (because of excessive caution and focus on compliance).

Despite the Push to Accept Additional Risk, There Is Still a Cultural Aversion to Risk in the Australian, Canadian, and British Budgeting Processes

In Australia, stakeholders seek to spend within limits while adhering to the annual budget, which is intuitively prudent but could also limit agility. The cultural aversion to acquisition risk within Defence lengthens review times and holds up funds that could be spent on other projects.

Canada's political structure does not allow parliament to drastically change funding for departments, including DND, beyond what has been requested. Canada's political culture means that there is typically not much appetite for large increases in DND's spending in any given year.

As in DoD, the MoD is experimenting with new ways to encourage innovation, including a new dedicated Innovation Fund. However, these strategies have not alleviated the enduring challenge of a risk-averse MoD culture.

Applicability of Selected Allied and Partner Nation Insights to DoD's PPBE System

The Commission on PPBE Reform is looking for potential lessons from the PPBE-like systems of selected allied and partner nations to improve DoD's PPBE System. There are notable differences between the United States and the selected allies and partners in terms of political systems, population sizes, industrial bases, workforce sizes, and military expenditures; how-

---

[34] MoD, 2019b, p. 6.

[35] MoD, 2019b, p. 11.

[36] UK subject-matter expert, interview with the authors, November 2022.

ever, we found that, despite these differences, there are similarities in how all four countries generally approach defense resource management:

- Many decisionmakers and stakeholders are involved throughout the complicated defense resource allocation processes.
- Strategic planning is a key input that is used to explicitly connect priorities to how much funding is spent to address military threats.
- Ongoing discussions are held between defense departments and decisionmakers who control the "power of the purse" to justify how forces and programs will use the funding.
- Defense departments receive and spend funding according to agreed-on appropriations rules and then use certain mechanisms if plans change to move or carry over funding.
- Oversight is a key mechanism for making sure what is budgeted is appropriately spent.

The United States provides needed capabilities to Australia, Canada, and the UK. This dependence arises from the capacity of the U.S. industrial base and the technological edge of its systems, but it also arises from the high priority that these countries place on allied interoperability. Given the interdependencies that exist, the Commission on PPBE Reform may want to consider the consequences of potential changes to DoD's PPBE System for countries with some level of dependence on U.S. FMS.

Although the political systems of the allies and partners described in this report appear to offer easier ways to pass a defense budget with stronger executive branches, the U.S. system of government offers some of its own intended benefits of involving strong voices from both the executive and legislative branches. The diversity of thought can help ensure that both the majority and minority parties have some input in spending priorities. At the same time, this system can cause gridlock yearly through continuing resolutions and potential government shutdowns that allies do not endure. Continuing resolutions have been criticized for the inefficiency that they impose on DoD; the increased need for advance—or even crisis—planning; and the rush to spend when one-year funds are, at long last, available. Although DoD expects and prepares for annual continuing resolutions, the commission may want to consider alternatives for mitigating the consequences of these annual disruptions in resource allocations.

All three countries examined exert some level of oversight over defense spending. In all cases, a balance is needed between the necessary oversight and the necessary flexibility to support innovation in response to emerging priorities. All the cases demonstrated ways in which flexibility is afforded through various mechanisms. Although none is a magic bullet, certain allied mechanisms could help improve DoD practices. Of particular relevance are those mechanisms that provide extra budget surety for major multiyear investments as opposed to reevaluating them every year.

For example, the UK's PB&F system benefits from multiannual spending plans, programs, and contracts. The MoD can sign decade-long portfolio management agreements with UK firms to provide long-term certainty. The PB&F system also allows for advance funding early in a budget year to ensure continuous government operations, thereby avoiding the possibility—and cost—of a shutdown. Likewise, Australia's defense planning, programming, and budgeting processes provide a high level of certainty for the development and operationalization of major military capabilities. These farsighted processes ensure a strong connection between strategy and resources, reduce prospects for the misuse of funds or inefficiency, and limit the risk of blocked funding from year to year.

The Commission on PPBE Reform will find many similarities across processes used in the United States, Australia, Canada, and the UK, but one particular similarity that is ingrained in resource planning will be very tough to change: The risk-averse resource planning culture across these countries will need to adapt to allow additional ways to innovate to counter emerging and future threats.

Summary of the Governance and Budgetary Systems of Selected Allied and Partner Nation Case Studies

Finally, we provide a summary of the governance and budgetary systems of the allied and partner nation case studies with those of the United States for comparison in Tables 5.1 through 5.10.[37]

Tables 5.1 and 5.2 show comparisons of the governance structures of the United States, Australia, Canada, and the UK. Tables 5.3 through 5.10 compare the planning, programming, budgeting, and execution processes of the United States, Australia, Canada and the UK.

---

[37] Information presented in these tables is derived from multiple sources and materials reviewed by the authors and cited elsewhere in this report. See the references list for full bibliographic details.

TABLE 5.1

Governance: U.S. and Comparative Nation Government Structures and Key Participants

Country	Structure of Government or Political System	Key Governing Bodies and Participants
United States	Federal presidential constitutional republic	• President of the United States • Office of Management and Budget (OMB) • Congress (House of Representatives and Senate) • U.S. Department of Defense (DoD) • Secretary of Defense and senior DoD leadership • Joint Chiefs of Staff
Australia	Federal parliamentary constitutional monarchy	• Prime minister • Governor-general • Parliament (House of Representatives and Senate) • Minister for Defence • Department of Defence
Canada	Federal parliamentary constitutional monarchy	• Prime minister • Governor general • Parliament (House of Commons and Senate) • Department of National Defence (DND) • Minister of Finance • Minister of National Defence • Deputy Minister of National Defence
UK	Unitary parliamentary constitutional monarchy	• Prime minister • Parliament (House of Commons and House of Lords) • Ministry of Defence (MoD) • Secretary of State for Defence • Permanent Under-Secretary of State for Defence

TABLE 5.2

Governance: U.S. and Comparative Nation Spending Controls and Decision Supports

Country	Control of Government Spending	Decision Support Systems
United States	Legislative review and approval of executive budget proposal	• Planning, Programming, Budgeting, and Execution (PPBE) System • Joint Capabilities Integration and Development System (JCIDS) • Defense Acquisition System (DAS)
Australia	Executive with legislative review and approval. Appropriations legislation must originate in the House of Representatives; Senate may reject but cannot amend.	One Defense Capability System (ODCS), including the following: • the Integrated Force Design Process, featuring a two-year cycling Defence Capability Assessment Program (DCAP) • the Integrated Investment Program (IIP), which documents planned future capability investments and informs the Portfolio Budget Statement, the proposed allocation of resources to outcomes • acquisition of approved IIP capability programs • sustainment and disposal of capability programs.
Canada	Executive with assessed limited influence of legislative review and approval	• Expenditure Management System • Defence Capabilities Board • Independent Review Panel for Defence Acquisition
UK	Executive with legislative review and approval	• Public Finance Management Cycle • Planning, Budgeting, and Forecasting (PB&F) • Defence Operating Model

TABLE 5.3

Planning: U.S. and Comparative Nation Inputs and Outputs

Country	Key Planning Inputs	Selected Planning Outputs
United States	• National Security Strategy • National Defense Strategy • National Military Strategy	• Chairman's Program Recommendations • Defense Planning Guidance • Fiscal Guidance
Australia	• 2016 Defence White Paper • 2017 Defence Industry Policy Statement • 2017 Strategy Framework • 2019 Defence Policy for Industry Participation • 2020 Defence Strategic Update • 2020 Force Structure Plan • 2023 Defence Strategic Review • Defence Planning Guidance / Chief of the Defence Force Preparedness Directive (Not available to the general public) • Other strategic plans and documents outlining planning and program requirements	• IIP for future capability investment
Canada	• 2017 Defence White Paper (*Strong, Secure, Engaged*) • 2018 Defence Plan, 2018–2023 • 2019 Defence Investment Plan • 2020 Defence Capabilities Blueprint (updated monthly) • 2022 Department of National Defence and Canadian Armed Forces Engagement Plan (released annually)	• Annual department plans to link DND strategic priorities and expected program results to the Main Estimates presented to parliament
UK	• Public Finance Management Cycle • PB&F • Defence Operating Model	• 2021 Defence Command Paper (*Defence in a Competitive Age*) aligns MoD priorities with the Integrated Review • 2021 Defence and Security Industrial Strategy

TABLE 5.4

Planning: U.S. and Comparative Nation Strategic Emphasis and Stakeholders

Country	Strategic Planning Emphasis	Planning Stakeholders
United States	2022 National Defense Strategy highlights four priorities: (1) defending the United States, "paced to the growing multi-domain threat posed by the [People's Republic of China (PRC)]"; (2) deterring "strategic attacks against the United States, Allies, and partners"; (3) deterring aggression and being prepared to "prevail in conflict when necessary" with priority placed first on the PRC "challenge in the Indo-Pacific region" and then "the Russia challenge in Europe"; and (4) "building a resilient Joint Force and defense ecosystem."	• Under Secretary of Defense for Policy (lead actor, produces Defense Planning Guidance) • President (National Security Strategy, Fiscal Guidance) • Secretary of Defense (National Defense Strategy, Fiscal Guidance at DoD level) • Chairman of the Joint Chiefs of Staff (CJCS) (National Military Strategy, Chairman's Program Recommendations)
Australia	2023 Defence Strategic Review emphasized a strategy of deterrence to deny an adversary freedom of action to militarily coerce Australia and to operate against Australia without being held at risk	• Strategic guidance generated by Department of Defence; approved by the Minister for Defence • IIP managed by the Vice Chief of the Defence Force, with input from stakeholders and joint strategic planning units, such as the Force Design Division
Canada	2017 white paper emphasized three components to Canadian national defense: (1) defense of national sovereignty through Canadian Armed Forces capable of assisting in response to natural disasters, search and rescue, and other emergencies; (2) defense of North America through partnership in NORAD with the United States; and (3) international engagements, including through peace support operations and peacekeeping.	• DND and supporting cabinet entities
UK	2021 Defence Command Paper emphasized seven primary goals of the MoD and the British Armed Forces: (1) defense of the UK and its overseas territories; (2) sustainment of UK nuclear deterrence capacity; (3) global influence projection; (4) execution of NATO responsibilities; (5) promotion of national prosperity; (6) peacekeeping contributions; and (7) supporting defense and intelligence-gathering capabilities of UK allies and partners.	• Prime Minister's Cabinet Office (Integrated Review) • MoD (Defense Command Paper and other strategic documents)

TABLE 5.5

Programming: U.S. and Comparative Nation Resource Allocations and Time Frames

Country	Resource Allocation Decisions	Programming Time Frames
United States	Documented in program objective memorandum (POM) developed by DoD components, reflecting a "systematic analysis of missions and objectives to be achieved, alternative methods of accomplishing them, and the effective allocation of the resources," and reviewed by the Director of Cost Assessment and Program Evaluation (CAPE)	• 5 years
Australia	Portfolio Budget Statement (as informed by the IIP) for the current fiscal year	Three-tiered funding stream that provides • current fiscal year funding • forward-looking estimates with a high degree of confidence for the next 3 fiscal years • provisional funding with a medium degree of confidence for the next 10 years, as articulated in the IIP and defense strategic guidance documents.
Canada	Government Expenditure Plan and Main Estimates allocate budget resources to departments and programs	• 3 years, as articulated in the Annual Department Plan
UK	• Main supply estimates (MEs) for the current fiscal year, based on spending limits set in the Integrated Review, and additional estimates for 10 years out as articulated in the MoD *Defence Equipment Plan*, which is updated annually • Supplementary supply estimates (SEs) allow the MoD to request additional resources, capital, or cash for the current fiscal year. • Excess votes—although discouraged—allow retroactive approval of overruns from a prior fiscal year, because government departments cannot legally spend more money than has been approved by parliament.	• 3–5 years, as articulated in the Integrated Review, which provides medium-term financial planning

TABLE 5.6

Programming: U.S. and Comparative Nation Stakeholders

Country	Programming Stakeholders
United States	• Director, CAPE (lead actor, provides analytic baseline to analyze POM produced by DoD components, leads program reviews, forecasts resource requirements, and updates the Future Years Defense Program [FYDP]) • DoD components (produce POM, document proposed resource requirements for programs over 5-year time span, which comprises the FYDP) • CJCS (assesses component POMs, provides chairman's program assessment reflecting the extent to which the military departments [MILDEPs] have satisfied combatant command [COCOM] requirements) • Deputy Secretary of Defense (adjudicates disputes through the Deputy's Management Action Groups) • Secretary of Defense (as needed, directs DoD components to execute Resource Management Decision memoranda to reflect decisionmaking during the programming and budget phases)
Australia	• At Department of Defence level, decisionmaking for resources made through Defence Committee (Defence Secretary, Chief of the Defence Force, Vice Chief of the Defence Force, Associate Defence Secretary, Chief Finance Officer) • Capability-related submissions reviewed by Minister for Finance–led National Security Investment Committee of Cabinet • Approved by National Security Committee (prime minister, deputy prime minister, Minister for Defence, Treasurer, Minister for Finance, other ministers when necessary)
Canada	• Treasury Board and Department of Finance (sets annual spending limits for federal agencies that are applicable to capital expenditures and determines the number of new projects funded) • Department of Finance, led by the Minister of Finance (drafts budget for presentation to parliament) • Minister of Finance and prime minister have approval authority. • Assistant Deputy Minister for Finance, DND and the DND Finance Group (prepares DND budget and liaises with Treasury Board Secretariat, Department of Finance, and other federal agencies) • Military service comptrollers
UK	• Component entities negotiate with MoD through demand signals; components program against required outputs. • MoD reviews and prioritizes proposed programs through a centralized process. • MoD Director General, Finance, working with the Deputy Chief of the Defense Staff for Military Capability (part of the MoD Financial and Military Capability team) • Supported by Director of Financial Planning and Scrutiny and the Assistant Chief of the Defence Staff for Capability and Force Design (part of the Financial and Military Capability team) • Process execution delegated to the Head of Defence Resources

TABLE 5.7

Budgeting: U.S. and Comparative Nation Time Frames and Major Categories

Country	Budget Approval Time Frames	Major Budget Categories
United States	Annual	• 5 categories: Military Personnel (MILPERS); Operation and Maintenance (O&M); Procurement; Research, Development, Test, and Evaluation (RDT&E); and Military Construction (MILCON)
Australia	• Annual, with separate appropriations bills for existing services and programs and for new programs • Accrual budgeting with budget request covering ongoing costs; associated funding cannot be carried over to the next fiscal year	• 5 categories: Workforce, Operations, Capability Acquisition Programs (including research and development), Capability Sustainment, and Operating Costs
Canada	Annual; disbursement of funds made through 3 supply periods, each reviewed and approved by parliament with Main Estimates and Supplementary Estimate A (i.e., spending not ready to be included in the Main Estimate at time of preparation) presented in first supply period. Supplementary Estimate B is presented in the second supply period, and Supplementary Estimate C (as needed) is presented in the third supply period.	• Various categories: *votes* for separate tranches of funding roughly correspond to DoD's colors of money. FY 2022–FY 2023 contained four votes for (1) operating expenditures; (2) capital expenditures, including major capability programs and infrastructure projects; (3) grants and contributions, including payments to NATO and funding for partner-nation military programs; and (4) payments for long-term disability and life insurance plans for Canadian Armed Forces Members. • Main Estimates also categorize spending by purpose. FY 2022–FY 2023 purpose categorizations include such areas as (1) ready forces, (2) capability procurement, (3) future force design, and (4) operations.
UK	Annual	• 8 categories, as split by the MoD for its internal PPBE-like process, corresponding to 8 main MoD organizations, central oversight to promote jointness • Budgets divided into commodity blocks (capital departmental expenditure limit for investment, resource departmental expenditure limit for current costs, etc.) and by activity (personnel, etc.)

TABLE 5.8

Budgeting: Selected U.S. and Comparative Nation Stakeholders

Country	Selected Budgeting Stakeholders
United States	DoD • Under Secretary of Defense (Comptroller) • DoD components and COCOMs Executive Branch • OMB Congress • House Budget Committee • Senate Budget Committee • House Appropriations Committee (Defense Subcommittee) • Senate Appropriations Committee (Defense Subcommittee) • House Armed Services Committee • Senate Armed Services Committee
Australia	Department of Defence management: • Vice Chief of the Defence Force • Associate Secretary of Defence • Investment Committee (chaired by the Vice Chief of the Defence Force; makes departmental decisions associated with execution of the IIP) • Capability managers (senior military officials, Chief Defence Scientist, Chief Information Officer (CIO), and the Deputy Secretary Security and Estate) and lead delivery groups. Decisions are ultimately the responsibility of the civilian executive government (prime minister, cabinet).
Canada	• Treasury Board and Department of Finance • Assistant Deputy Minister for Finance, DND and the DND Finance Group • Military service comptrollers
UK	• His Majesty's (HM) Treasury sets annual limits on net spending. • MoD drafts and presents MEs and SEs to parliament at different points within the fiscal year cycle, in close coordination with HM Treasury. • House of Commons Defence Select Committee examines MEs; parliament votes on MEs and SEs. • MoD Director General, Finance, working with the Deputy Chief of the Defense Staff for Military Capability (part of the MoD Financial and Military Capability team) • Supported by Director of Financial Planning and Scrutiny and the Assistant Chief of the Defence Staff for Capability and Force Design (part of the Financial and Military Capability team) • Process execution delegated to the Head of Defence Resources

TABLE 5.9

Execution: U.S. and Comparative Nation Budgetary Flexibilities and Reprogramming

Country	Budgetary Flexibilities and Reprogramming
United States	• Funding availability varies by account type; multiyear or no-year appropriations for limited programs as authorized by Congress • Limited carryover authority in accordance with OMB Circular A-11 • Reprogramming as authorized; four defined categories of reprogramming actions, including prior-approval reprogramming actions—increasing procurement quantity of a major end item, establishing a new program, etc.—which require approval from congressional defense committees • Transfers as authorized through general and special transfer authorities, typically provided in defense authorization and appropriations acts
Australia	• Ten-year indicative baseline for defense spending (except operating costs) provides budgetary certainty entering into each new fiscal year. • IIP includes approved capability development programs—for which funding does not expire—and unapproved programs that can be accelerated or delayed as needs arise or change to reallocate funds through biannual review process overseen by the Vice Chief of the Defence Force, including between services and for new projects • The IIP is 20% overprogrammed for acquisition to manage risks of underachievement or overexpenditure relative to the acquisition budget. • Funding for operations, sustainment, and personnel is separate from the IIP. • Capability managers have a high degree of flexibility for spending allocated operating funds; responsible for achieving outcomes articulated in the Portfolio Budget Statement.
Canada	• Organizations can transfer funds within a vote from one program to another without parliament's approval. • Organizations do need parliament's approval to transfer funds between votes. • Canadian federal agencies are allowed to carry forward a portion of unspent funds for a fiscal year—typically up to 5% of operating expenditures and 20% of capital expenditures. • Government can authorize continued spending at prior-year levels if a budget has not been passed by parliament by the beginning of the fiscal year. • Special warrants can be issued to fund continued normal government operations if a government falls and an election is called before a budget can be passed; this can also be used on a short-term basis to avoid the need for a parliament vote on funding. • Interim supply bill for a new fiscal year typically presented and voted on in third supply period of prior fiscal year to allow continued government operations, as budget and Main Estimates are introduced close to the beginning of a new fiscal year.
UK	• Defense operations funded separately through HM Treasury or (in certain circumstances) UK Integrated Security Fund (as managed by the Cabinet Office's Joint Funds Unit • Already voted funding can be moved within top-line budget programs with HM Treasury approval, provided they remain in the same commodity block. • MoD funds can also be directly transferred between programs within a departmental expenditure limit or annual managed expenditure in a process known as *virement*, subject to restrictions. • Additional funding for one or more top-line budget programs can be requested from parliament as an SE. • Portions of budget subject to highest degree of fluctuation treated as annual managed expenditures (with operations covered through HM Treasury and/or UK Integrated Security Fund); MoD can request additional funds from HM Treasury to support urgent and unanticipated needs.

TABLE 5.10

Execution: U.S. and Comparative Nation Assessment

Country	Key Stakeholders in Execution Assessment
United States	<ul><li>Under Secretary of Defense (Comptroller)</li><li>DoD component comptrollers and financial managers</li><li>Department of the Treasury</li><li>Government Accountability Office</li><li>OMB</li><li>Defense Finance and Accounting Service</li></ul>
Australia	<ul><li>National Audit Office</li><li>Finance regulations within Department of Defence and the public service</li><li>Defence Finance Policy Framework</li><li>Annual Performance Statement; submitted in October of the year following defense appropriation by the prime minister and cabinet</li><li>Portfolio Additional Estimates Statement; reflects budget appropriations and changes between budgets</li></ul>
Canada	<ul><li>Auditor General</li><li>Parliamentary Budget Office</li><li>DND internal Review Services division</li></ul>
UK	<ul><li>National Audit Office</li><li>Comptroller and Auditor General</li><li>HM Treasury (approval required for any MoD expenditure above £600 million, monthly and annual reporting from MoD on actual and forecasted spending, etc.)</li><li>House of Commons Public Accounts Committee</li></ul>

Abbreviations

ADF	Australian Defence Force
AME	annual managed expenditure
ASCA	Australian Strategic Capabilities Accelerator
AUKUS	Australia–United Kingdom–United States
CAAS	Cost Assurance and Analysis Service
CAF	Canadian Armed Forces
CAPE	Cost Assessment and Program Evaluation
CDF	Chief of the Defence Force
CFO	Chief Finance Officer
COCOM	combatant command
COVID-19	coronavirus disease 2019
CSSF	Conflict, Stability and Security Fund
DCAP	Defence Capability Assessment Program
DCB	Defence Capabilities Board
DE&S	Defence Equipment and Support
DEL	departmental expenditure limit
DIH	Defence Innovation Hub
DLOD	defense line of development
DND	Canadian Department of National Defence
DoD	U.S. Department of Defense
DoDD	Department of Defense Directive
DOTMLPF	doctrine, organization, training, materiel, leadership and education, personnel, and facilities
DP	departmental plan
DRR	departmental results report
DSIS	Defence and Security Industrial Strategy
DSR	Defense Strategic Review
FIC	fundamental input to capability
FLC	Front Line Command
FMS	Foreign Military Sales
FY	fiscal year
FYDP	Future Years Defense Program
GDP	gross domestic product

IIP	Integrated Investment Program
IPA	Infrastructure and Projects Authority
IRPDA	Independent Review Panel for Defence Acquisition
JSP	joint service publication
ME	main supply estimate
MoD	UK Ministry of Defence
MP	member of parliament
NAO	UK National Audit Office
NATO	North Atlantic Treaty Organization
NDS	National Defense Strategy
NORAD	North American Aerospace Defense Command
NSC	National Security Committee of Cabinet
NTIB	National Technology and Industrial Base
O&M	operation and maintenance
ODCS	One Defense Capability System
OMB	Office of Management and Budget
OSCAR	Online System for Central Accounting and Reporting
OSD	Office of the Secretary of Defense
PB&F	Planning, Budgeting, and Forecasting
PPBE	Planning, Programming, Budgeting, and Execution
PPBS	Planning, Programming, and Budgeting System
PRC	People's Republic of China
R&D	research and development
RDT&E	research, development, test, and evaluation
SCN	Shipbuilding and Conversion, Navy
SE	supplementary supply estimate
SIPRI	Stockholm International Peace Research Institute
SSRO	Single Source Regulations Office
TLB	top-level budget
TRL	technology readiness level
UCR	urgent capability requirement
UK	United Kingdom
UKISF	United Kingdom Integrated Security Fund
VCDF	Vice Chief of the Defence Force

References

Armit, Amelita A., "An Overview of the Canadian Budget Process," paper presented to a Roundtable on State Financial Control at the Ninth St. Petersburg International Economic Forum, Moscow, Russia, June 2005.

Australian Academy of Science, *2023–24 Pre-Budget Submission*, January 2023.

"Australian Defence Projects Billions over Budget, Decades Late," Reuters, October 9, 2022.

Australian Department of Defence, "Budgets," webpage, undated-a. As of September 1, 2022: https://www.defence.gov.au/about/information-disclosures/budgets

Australian Department of Defence, "Sovereign Industrial Base Capability Priorities and Plans," webpage, undated-b. As of September 9, 2022: https://www.defence.gov.au/business-industry/capability-plans/ sovereign-industrial-capability-priorities

Australian Department of Defence, "Strategic Planning," webpage, undated-c. As of February 6, 2023: https://www.defence.gov.au/about/strategic-planning

Australian Department of Defence, *Australian Defence*, November 1976.

Australian Department of Defence, *2016 Defence Industry Policy Statement*, 2016a.

Australian Department of Defence, *2016 Defence White Paper*, 2016b.

Australian Department of Defence, "Part 3: Government Direction," *The Strategy Framework*, 2017.

Australian Department of Defence, *2018 Defence Industrial Capability Plan*, 2018.

Australian Department of Defence, *Defence Policy for Industry Participation*, 2019.

Australian Department of Defence, *2020 Defence Strategic Update*, 2020a.

Australian Department of Defence, *2020 Force Structure Plan*, 2020b.

Australian Department of Defence, *Department of Defence Annual Report 2019–20*, 2020c.

Australian Department of Defence, *Defence Capability Manual*, version 1.1, December 3, 2021a.

Australian Department of Defence, *Force Design Guidance*, version 1.1, December 3, 2021b.

Australian Department of Defence, *Defence Corporate Plan: 2022–2026*, August 2022a.

Australian Department of Defence, *Portfolio Budget Statements 2022–23: Defence Portfolio— Budget Initiatives and Explanations of Appropriations Specified by Outcomes and Programs by Entity: Australian Signals Directorate*, October 2022b.

Australian Department of Defence, *Portfolio Budget Statements 2022–23: Defence Portfolio— Entity Resources and Planned Performance*, October 2022c.

Australian Department of Defence, *National Defence: Defence Strategic Review*, 2023.

Australian Government, "Portfolio Budget Statements," webpage, undated. As of October 10, 2023: https://budget.gov.au/content/pbs/index.htm

Australian National Audit Office, *Defence's Administration of the Integrated Investment Program*, Auditor-General Report No. 7 2022–23, 2022.

Australian Parliament, "Double Dissolutions," webpage, undated. As of September 5, 2022: https://www.aph.gov.au/About_Parliament/House_of_Representatives/ Powers_practice_and_procedure/Practice7/HTML/Chapter13/Double_dissolutions

Australian Parliament, Joint Standing Committee on Foreign Affairs, Defence and Trade, *Funding Australia's Defence*, May 8, 1998.

Australian Parliamentary Budget Office, "Overview of the Budget Process," fact sheet, 2022.

Australian Parliamentary Education Office, "Question Time in the House of Representatives," fact sheet, 2022.

Australian Senate, "Consideration of Estimates by the Senate's Legislation Committees," Senate Brief No. 5, January 2023.

Barnes, Andre, *Appointment of Officers of Parliament*, Library of Parliament, Publication No. 2009-21-E, August 19, 2021.

Berthiaume, Lee, "Auditors Call Out National Defence for Poor Oversight on Spending," Global News, June 14, 2020.

Black, James, Richard Flint, Ruth Harris, Katerina Galai, Pauline Paillé, Fiona Quimbre, and Jessica Plumridge, *Understanding the Value of Defence: Towards a Defence Value Proposition for the UK*, RAND Corporation, RR-A529-1, 2021. As of March 6, 2023: https://www.rand.org/pubs/research_reports/RRA529-1.html

Boroush, Mark, and Ledia Guci, *Science and Engineering Indicators 2022: Research and Development—U.S. Trends and International Comparisons*, National Science Board, April 28, 2022.

Boyce, J., N. Tay, and C. Row, "Consideration of Enabling and Enterprising Functions Within Defence Force Design," *Proceedings of the 24th International Congress on Modelling and Simulation*, December 2021.

Brangwin, Nicole, and David Watt, *The State of Australia's Defence: A Quick Guide*, Australian Parliament, July 27, 2022.

Brennan, Tim, Hazel Ferguson, and Ian Zhou, "Science and Research," *Budget Review 2022–23*, Australian Parliament, April 2022.

Brodie, Bernard, *Strategy in the Missile Age*, RAND Corporation, CB-137-1, 1959. As of April 21, 2023: https://www.rand.org/pubs/commercial_books/CB137-1.html

Busby, Carleigh, Albert Kho, and Christopher E. Penney, *The Life Cycle Cost of the Canadian Surface Combatants: A Fiscal Analysis*, Office of the Parliamentary Budget Officer, October 27, 2022.

Campbell, Meagan, "How Canada Avoids U.S.-Style Government Shutdowns," *Maclean's*, January 23, 2018.

Canadian Department of National Defence, *Strong, Secure, Engaged: Canada's Defence Policy*, 2017.

Canadian Department of National Defence, *Defence Investment Plan 2018: Ensuring the Canadian Armed Forces Is Well-Equipped and Well-Supported*, 2018a.

Canadian Department of National Defence, *Defence Plan: 2018–2023*, 2018b.

Canadian Department of National Defence, "Defence Purchases and Upgrades Process," webpage, updated September 10, 2018c. As of March 2, 2023:
https://www.canada.ca/en/department-national-defence/services/procurement/defence-purchases-upgrades-process.html

Canadian Department of National Defence, "Mandate of National Defence and the Canadian Armed Forces," webpage, updated September 24, 2018d. As of March 2, 2023:
https://www.canada.ca/en/department-national-defence/corporate/mandate.html

Canadian Department of National Defence, *Defence Investment Plan 2018: Annual Update 2019—Ensuring the Canadian Armed Forces Is Well-Equipped and Well-Supported*, 2019.

Canadian Department of National Defence, "Defence Capabilities Blueprint," webpage, updated January 9, 2020a. As of March 2, 2023:
http://dgpaapp.forces.gc.ca/en/defence-capabilities-blueprint/index.asp

Canadian Department of National Defence, "Supplementary Estimates (A)—National Defence," webpage, updated April 7, 2020b. As of March 2, 2023:
https://www.canada.ca/en/department-national-defence/corporate/reports-publications/proactive-disclosure/cow-estimates-a-2019-20/supp-estimate-a.html

Canadian Department of National Defence, *Department of National Defence and Canadian Armed Forces, 2022–2023: Departmental Plan*, 2022a.

Canadian Department of National Defence, "Current Operations and Joint Military Exercises List," webpage, updated January 13, 2022b. As of March 2, 2023:
https://www.canada.ca/en/department-national-defence/services/operations/military-operations/current-operations/list.html

Canadian Department of National Defence, "DND/CAF Organizational Chart," webpage, February 23, 2022c. As of March 2, 2023:
https://www.canada.ca/en/department-national-defence/corporate/reports-publications/transition-materials/mnd-transition-material-2021-dnd/tab6-dnd-caf-org-chart.html

Canadian Department of National Defence, "Defence Capabilities Blueprint," webpage, updated December 1, 2022d. As of May 9, 2023:
http://dgpaapp.forces.gc.ca/en/defence-capabilities-blueprint/index.asp

Canadian Department of National Defence, "Defence Investment Plan 2018," webpage, updated December 9, 2022e. As of March 2, 2023:
https://www.canada.ca/en/department-national-defence/corporate/reports-publications/defence-investment-plan-2018.html

Chapman, Bert, "The Geopolitics of Canadian Defense White Papers: Lofty Rhetoric and Limited Results," *Geopolitics, History, and International Relations*, Vol. 11, No. 1, 2019.

Chewning, Eric, Will Gangware, Jess Harrington, and Dale Swartz, *How Will US Funding for Defense Technology Innovation Evolve?* McKinsey and Company, November 4, 2022.

Clarke, Michael, "Army, Navy and RAF: Winners and Losers of Defence's Transformation," webpage, Forces.net, March 22, 2021. As of March 6, 2023:
https://www.forces.net/comment/pain-and-promises-transforming-armed-forces-integrated-review

Congressional Research Service, *A Defense Budget Primer*, RL30002, December 9, 1998.

Conservative Party, *Get Brexit Done, Unleash Britain's Potential: Conservative and Unionist Party Manifesto 2019*, 2019.

Cook, Cynthia R., Emma Westerman, Megan McKernan, Badreddine Ahtchi, Gordon T. Lee, Jenny Oberholtzer, Douglas Shontz, and Jerry M. Sollinger, *Contestability Frameworks: An International Horizon Scan*, RAND Corporation, RR-1372-AUS, 2016. As of February 6, 2023: https://www.rand.org/pubs/research_reports/RR1372.html

Davies, Gareth, *Improving the Performance of Major Equipment Contracts*, National Audit Office, June 22, 2021.

Department of Defense Directive 7045.14, *The Planning, Programming, Budgeting, and Execution (PPBE) Process*, August 29, 2017.

Deraspe, Raphaëlle, *Funding New Government Initiatives: From Announcement to Money Allocation*, Library of Parliament, Publication No. 2021-32-E, October 7, 2021.

DND—*See* Canadian Department of National Defence.

DoD—*See* U.S. Department of Defense.

DoDD—*See* Department of Defense Directive.

Enthoven, Alain C., and K. Wayne Smith, *How Much Is Enough? Shaping the Defense Program, 1961–1969*, RAND Corporation, CB-403, 1971.

Ferguson, Gregor, "Peever Review Could Transform Defence Innovation and Acquisition," *The Australian*, October 30, 2021.

Forsey, Eugene A., *How Canadians Govern Themselves*, 10th ed., Library of Parliament, March 2020.

Fusaro, Paola, Nicolas Jouan, Lucia Retter, and Benedict Wilkinson, *Science and Technology as a Tool of Power: An Appraisal*, RAND Corporation, PE-A2391-1, November 2022. As of March 6, 2023: https://www.rand.org/pubs/perspectives/PEA2391-1.html

Government of Canada, "The Reporting Cycle for Government Expenditures," webpage, updated June 17, 2010. As of March 2, 2023: https://www.canada.ca/en/treasury-board-secretariat/services/planned-government-spending/expenditure-management-system/reporting-cycle.html

Government of Canada, "The Canada-U.S. Defence Relationship," webpage, updated April 25, 2014. As of March 2, 2023: https://www.canada.ca/en/news/archive/2014/04/canada-defence-relationship.html

Government of Canada, "Governor General's Special Warrants," webpage, updated October 19, 2015. As of March 2, 2023: https://www.canada.ca/en/treasury-board-secretariat/services/planned-government-spending/governor-general-special-warrants.html

Government of Canada, "Policy on Results," webpage, updated July 1, 2016a. As of March 2, 2023: https://www.tbs-sct.canada.ca/pol/doc-eng.aspx?id=31300

Government of Canada, "Departmental Results Reports," webpage, updated November 21, 2016b. As of March 2, 2023: https://www.canada.ca/en/treasury-board-secretariat/services/departmental-performance-reports.html

Government of Canada, "Defence Procurement Strategy," webpage, updated November 3, 2021a. As of March 2, 2023: https://www.tpsgc-pwgsc.gc.ca/app-acq/amd-dp/samd-dps/index-eng.html

Government of Canada, "Defence and Marine Procurement," webpage, updated November 23, 2021b. As of March 2, 2023:
https://www.tpsgc-pwgsc.gc.ca/app-acq/amd-dp/index-eng.html

Government of Canada, "Chapter 5: Canada's Leadership in the World," in *Budget 2022: A Plan to Grow Our Economy and Make Life More Affordable*, 2022a, pp. 131–146.

Government of Canada, "Main Estimates—2022–23 Estimates," webpage, updated March 1, 2022b. As of January 22, 2023:
https://www.canada.ca/en/treasury-board-secretariat/services/planned-government-spending/government-expenditure-plan-main-estimates/2022-23-estimates/main-estimates.html

Government of Canada, GC InfoBase, "Infographic for National Defence: Results," webpage, December 2, 2022. As of December 5, 2022:
https://www.tbs-sct.canada.ca/ems-sgd/edb-bdd/index-eng.html#infographic/dept/133/results/.-.-(panel_key.-.-'drr_summary)

Government of Canada, Global Affairs Canada, "Planning and Performance," webpage, updated March 3, 2022. As of March 2, 2023:
https://www.international.gc.ca/gac-amc/publications/plans/index.aspx?lang=eng

Government of Canada, Justice Laws Website, "The Constitution Acts, 1867 to 1982: Section VI. Distribution of Legislative Powers—Powers of the Parliament," webpage, updated February 17, 2023. As of March 2, 2023:
https://laws-lois.justice.gc.ca/eng/const/page-3.html#h-18

Government of Canada, Privy Council Office, *Guide for Parliamentary Secretaries*, December 2015.

Government Risk Protection and the Risk Centre of Excellence, *The Orange Book: Management of Risk—Principles and Concepts*, 2020.

Greenwalt, William, and Dan Patt, *Competing in Time: Ensuring Capability Advantage and Mission Success Through Adaptable Resource Allocation*, Hudson Institute, February 2021.

Hellyer, Marcus, "The Real Costs of Australia's Defence Budget 'Blowout,'" webpage, The Strategist, October 18, 2022. As of October 31, 2022:
https://www.aspistrategist.org.au/the-real-costs-of-australias-defence-budget-blowout

Hellyer, Marcus, and Ben Stevens, *The Cost of Defence: ASPI Defence Budget Brief 2022–2023*, Australian Strategic Policy Institute, June 2022.

Hitch, Charles J., and Roland N. McKean, *The Economics of Defense in the Nuclear Age*, RAND Corporation, R-346, 1960. As of April 21, 2023:
https://www.rand.org/pubs/reports/R346.html

HM Treasury, *The Aqua Book: Guidance on Producing Quality Analysis for Government*, March 2015.

HM Treasury, *The Public Value Framework: With Supplementary Guidance*, March 2019.

HM Treasury, *Magenta Book: Central Government Guidance on Evaluation*, March 2020.

HM Treasury, "Autumn Budget and Spending Review 2021 Representations," webpage, updated September 7, 2021. As of March 6, 2023:
https://www.gov.uk/government/publications/autumn-budget-and-spending-review-2021-representations

HM Treasury, "Contingencies Fund Account 2021 to 2022," webpage, June 16, 2022a. As of March 6, 2023:
https://www.gov.uk/government/publications/contingencies-fund-account-2021-to-2022/contingencies-fund-account-2021-22

HM Treasury, "The Green Book," webpage, updated November 18, 2022b. As of March 6, 2023:
https://www.gov.uk/government/publications/the-green-book-appraisal-and-evaluation-in-central-governent/the-green-book-2020

Hurst, Daniel, "Defence Projects Suffer $6.5bn Cost Blowout as Marles Promises More Scrutiny in Future," *The Guardian*, October 9, 2022.

Insinna, Valerie, "Former US Air Force Acquisition Czar Could Help the UK Build Its Future Fighter," *Defense News*, September 14, 2021.

IPA—*See* UK Infrastructure and Projects Authority.

Jackson, Lewis, "Australia's Nuclear Submarine Plan to Cost up to $245 billion by 2055—Defence Official," Reuters, March 14, 2023.

Johnson, Robin, "UK: Ministry of Defence (MoD) Proposals for GOCO Shelved in Favour of DE&S Plus Variant," Eversheds Sutherland International, December 11, 2013.

Johnstone, Richard, "First Class Delivery: What the MoD Team Renewing the UK's Nuclear Submarines Learnt from the Olympics and Crossrail," *Civil Service World*, May 29, 2018.

Kincaid, Bill, *Changing the Dinosaur's Spots: The Battle to Reform UK Defence Acquisition*, Royal United Services Institute, 2008.

Kirk-Wade, Esme, *UK Defence Expenditure*, House of Commons Library, April 6, 2022.

Lagassé, Philippe, "Improving Parliamentary Scrutiny of Defence," *Canadian Military Journal*, Vol. 22, No. 3, Summer 2022.

Lang, Eugene, "The Shelf Life of Defence White Papers," *Policy Options*, June 23, 2017.

"Liz Truss Defence Spending to Cost £157bn, Says Report," BBC, September 2, 2022.

Lucas, Rebecca, Lucia Retter, and Benedict Wilkinson, *Realising the Promise of the Defence and Security Industrial Strategy in R&D and Exports*, RAND Corporation, PE-A2392-1, November 2022. As of March 6, 2023:
https://www.rand.org/pubs/perspectives/PEA2392-1.html

MacLennan, Leah, "Warning of Defence Shipbuilding Skills Shortage Amid Uncertainty over Local Submarine Build," Australian Broadcasting Corporation News, May 10, 2022.

Marles, Richard, "Television Interview, ABC News Broadcast," transcript, March 15, 2023.

Marles, Richard, and Pat Conroy, "Government Announces Most Significant Reshaping of Defence Innovation in Decades to Boost National Security," press release, April 28, 2023.

Massola, James, "Defence Facing a 'Personnel Crisis' with Thousands More Uniformed Members Needed," *Sydney Morning Herald*, November 14, 2022.

Mazarr, Michael J., *The Societal Foundations of National Competitiveness*, RAND Corporation, RR-A499-1, 2022. As of April 21, 2023:
https://www.rand.org/pubs/research_reports/RRA499-1.html

MBDA Missile Systems, "The Portfolio Management Agreement," webpage, undated. As of March 6, 2023:
https://www.mbda-systems.com/about-us/mission-strategy/team-complex-weapons/the-portfolio-management-agreement

McGarry, Brendan W., *Defense Primer: Planning, Programming, Budgeting and Execution (PPBE) Process*, Congressional Research Service, IF10429, January 27, 2020.

McGarry, Brendan W., *DOD Planning, Programming, Budgeting, and Execution: Overview and Selected Issues for Congress*, Congressional Research Service, R47178, July 11, 2022.

McKernan, Megan, Stephanie Young, Ryan Consaul, Michael Simpson, Sarah W. Denton, Anthony Vassalo, William Shelton, Devon Hill, Raphael S. Cohen, John P. Godges, Heidi Peters, and Lauren Skrabala, *Planning, Programming, Budgeting, and Execution in Comparative Organizations*: Vol. 3, *Case Studies of Selected Non-DoD Federal Agencies*, RAND Corporation, RR-A2195-3, 2024. As of January 12, 2024:
www.rand.org/pubs/research_reports/RRA2195-3

McKernan, Megan, Stephanie Young, Timothy R. Heath, Dara Massicot, Andrew Dowse, Devon Hill, James Black, Ryan Consaul, Michael Simpson, Sarah W. Denton, Anthony Vassalo, Ivana Ke, Mark Stalczynski, Benjamin J. Sacks, Austin Wyatt, Jade Yeung, Nicolas Jouan, Yuliya Shokh, William Shelton, Raphael S. Cohen, John P. Godges, Heidi Peters, and Lauren Skrabala, *Planning, Programming, Budgeting, and Execution in Comparative Organizations*: Vol. 4, *Executive Summary*, RAND Corporation, RR-A2195-4, 2024. As of January 12, 2024:
www.rand.org/pubs/research_reports/RRA2195-4

McKernan, Megan, Stephanie Young, Timothy R. Heath, Dara Massicot, Mark Stalczynski, Ivana Ke, Raphael S. Cohen, John P. Godges, Heidi Peters, and Lauren Skrabala, *Planning, Programming, Budgeting, and Execution in Comparative Organizations*: Vol. 1, *Case Studies of China and Russia*, RAND Corporation, RR-A2195-1, 2024. As of January 12, 2024:
www.rand.org/pubs/research_reports/RRA2195-1

Meyer, Peter J., and Ian F. Fergusson, *Canada-U.S. Relations*, Congressional Research Service, No. 96-397, February 10, 2021.

MoD—*See* UK Ministry of Defence.

Morah, Chizoba, "Accrual Accounting vs. Cash Basis Accounting: What's the Difference?" webpage, Investopedia, March 19, 2023. As of March 23, 2023:
https://www.investopedia.com/ask/answers/09/accrual-accounting.asp

NAO—*See* UK National Audit Office.

Office of the Auditor General of Canada, "The Auditor General's Observations on the Government of Canada's 2018–2019 Consolidated Financial Statements," webpage, undated. As of March 2, 2023:
https://www.oag-bvg.gc.ca/internet/English/oag-bvg_e_43438.html

Office of the Auditor General of Canada, *Report 3: Supplying the Canadian Armed Forces—National Defence: Independent Auditor's Report*, Spring 2020.

Office of the Auditor General of Canada, *Report 2: National Shipbuilding Strategy: Independent Auditor's Report*, 2021.

Parliament of Canada, House Standing Committee on National Defence, "Meetings, 44th Parliament, 1st Session," webpage, undated. As of March 2, 2023:
https://www.ourcommons.ca/Committees/en/NDDN/Meetings

Parliament of Canada, House Standing Committee on National Defence, "Defence Spending Budget 2022," webpage, updated April 27, 2022. As of March 2, 2023:
https://www.canada.ca/en/department-national-defence/corporate/reports-publications/
proactive-disclosure/nddn-27-april-2022/defence-spending.html

Payne, Sebastian, and Sylvia Pfeifer, "Sunak Quiet on Defence Budget as He Signs Off on £4.2bn Frigate Contract," *Financial Times*, November 14, 2022.

Penney, Christopher E., *Canada's Military Expenditure and the NATO 2% Spending Target*, Office of the Parliamentary Budget Officer, June 9, 2022.

Penney, Christopher E., and Albert Kho, *Planned Capital Spending Under Strong, Secure, Engaged—Canada's Defence Policy: 2022 Update*, Office of the Parliamentary Budget Officer, March 11, 2022.

Perry, Dave, *A Primer on Recent Canadian Defence Budgeting Trends and Implications*, School of Public Policy Research Papers, University of Calgary, Vol. 8, No. 15, April 2015.

Perry, David, "DND Spending: A View from the Outside," briefing slides, Canadian Global Affairs Institute, Executive Leaders Program, September 12, 2019.

Perry, David, "Canadian Defence Budgeting," in Thomas Juneau, Philippe Lagassé, and Srdjan Vucetic, eds., *Canadian Defence Policy in Theory and Practice*, Palgrave Macmillan, 2020.

Pu, Shaowei, and Alex Smith, *The Parliamentary Financial Cycle*, Library of Parliament, Publication No. 2015-41-E, September 24, 2021.

Public Law 117-81, National Defense Authorization Act for Fiscal Year 2022, December 27, 2021.

Retter, Lucia, James Black, and Theodora Ogden, *Realising the Ambitions of the UK's Defence Space Strategy: Factors Shaping Implementation to 2030*, RAND Corporation, RR-A1186-1, 2022. As of March 6, 2023:
https://www.rand.org/pubs/research_reports/RRA1186-1.html

Retter, Lucia, Julia Muravska, Ben Williams, and James Black, *Persistent Challenges in UK Defence Acquisition*, RAND Corporation, RR-A1174-1, 2021. As of March 6, 2023:
https://www.rand.org/pubs/research_reports/RRA1174-1.html

Sabbagh, Dan, "Ben Wallace Steps Back from Liz Truss's 3% Defence Spending Target," *The Guardian*, November 10, 2022.

Sargent, John F., Jr., *U.S. Research and Development Funding and Performance: Fact Sheet*, Congressional Research Service, R44307, September 13, 2022.

SCAF, "About Us," webpage, undated. As of March 6, 2023:
https://scaf.org.uk/about-scaf/

Section 809 Panel, *Report of the Advisory Panel on Streamlining and Codifying Acquisition Regulations*, Vol. 2 of 3, June 2018.

Shapiro, Ariel, and Anne Marie Therrien-Tremblay, "Canada's Defence Policy Statements: Change and Continuity," HillNotes, Library of Parliament, September 22, 2022.

Sheldon, Tim, "Establishing a Project Controls Function at the UK Defence Equipment and Support Organisation," presentation at the Project Controls Expo 2017, London, November 16, 2017.

SIPRI—*See* Stockholm International Peace Research Institute.

Speciale, Stephen, and Wayne B. Sullivan II, "DoD Financial Management—More Money, More Problems," Defense Acquisition University, September 1, 2019.

Statistics Canada, "Departmental Plan," webpage, updated March 2, 2023. As of March 2, 2023:
https://www150.statcan.gc.ca/n1/en/catalogue/11-635-X

Stockholm International Peace Research Institute, "SIPRI Military Expenditure Database," homepage, undated. As of March 17, 2023:
https://milex.sipri.org/sipri

Stone, J. Craig, *Growing the Defence Budget: What Would Two Percent of GDP Look Like?* Canadian Global Affairs Institute, March 2017.

Strachan, Hew, and Ruth Harris, *The Utility of Military Force and Public Understanding in Today's Britain*, RAND Corporation, RR-A213-1, 2020. As of March 6, 2023:
https://www.rand.org/pubs/research_reports/RRA213-1.html

Thomas-Noone, Brendan, *Ebbing Opportunity: Australia and the US National Technology and Industrial Base*, United States Studies Centre, November 2019.

Trotter, Phillip, *The UK Central Government Financial Management System: A Guide for Stakeholders*, Institute of Chartered Accountants in England and Wales, November 2017.

Trudeau, Justin, "Delivering for Canadians Now," press release, March 22, 2022.

UK Cabinet Office, *Global Britain in a Competitive Age: The Integrated Review of Security, Defence, Development and Foreign Policy*, March 2021.

UK Cabinet Office, "PM Announces Major Defence Investment in Launch of Integrated Review Refresh," press release, March 13, 2023.

UK Conflict, Stability and Security Fund, "About Us," webpage, undated. As of March 6, 2023:
https://www.gov.uk/government/organisations/conflict-stability-and-security-fund/about

UK Foreign, Commonwealth and Development Office, "New Fund Announced to Support UK's National Security Priorities," press release, March 13, 2023.

UK Government Commercial Function, *Guide to Using the Social Value Model*, edition 1.1, December 3, 2020.

UK House of Commons, Defence Select Committee, "Memorandum for the Ministry of Defence: Supplementary Estimate 2021–22," March 2, 2021.

UK House of Commons, Foreign Affairs Committee, *Refreshing Our Approach? Updating the Integrated Review: Government Response to the Committee's Fifth Report—Fifth Special Report of Session 2022–23*, December 13, 2022.

UK House of Commons, Public Accounts Committee, "New Defence Money Potentially Lost in 'Funding Black Hole' at Centre of UK Defence Equipment Plan," March 16, 2021.

UK House of Commons, Public Accounts Committee, *Seventh Report of Session 2022–23: Armoured Vehicles: The Ajax Programme*, May 25, 2022.

UK Infrastructure and Projects Authority, *Infrastructure and Projects Authority Mandate*, January 2021.

UK Infrastructure and Projects Authority, *Annual Report on Major Projects, 2021–22*, July 2022.

UK Infrastructure and Projects Authority and UK Cabinet Office, "Infrastructure and Projects Authority: Assurance Review Toolkit," webpage, updated July 15, 2021. As of March 6, 2023:
https://www.gov.uk/government/collections/infrastructure-and-projects-authority-assurance-review-toolkit

UK Infrastructure and Projects Authority and UK Cabinet Office, "Major Projects Data," webpage, updated July 20, 2023. As of March 6, 2023:
https://www.gov.uk/government/collections/major-projects-data

UK Major Projects Authority, homepage, undated. As of March 6, 2023:
https://www.gov.uk/government/groups/major-projects-authority

UK Ministry of Defence, "About Us," webpage, undated-a. As of September 7, 2022:
https://www.gov.uk/government/organisations/ministry-of-defence/about

UK Ministry of Defence, "Defence Digital," webpage, undated-b. As of March 6, 2023:
https://www.gov.uk/government/groups/defence-digital

UK Ministry of Defence, "Defence and Security Accelerator," webpage, undated-c. As of March 6, 2023:
https://www.gov.uk/government/organisations/defence-and-security-accelerator

UK Ministry of Defence, "Defence Secretary Unveils Blueprint for Defence Reform," webpage, June 27, 2011. As of March 6, 2023:
https://www.gov.uk/government/news/defence-secretary-unveils-blueprint-for-defence-reform

UK Ministry of Defence, "Ministry of Defence Commercial," webpage, updated December 12, 2012. As of March 6, 2023:
https://www.gov.uk/guidance/ministry-of-defence-commercial

UK Ministry of Defence, *Investment Appraisal and Evaluation, Part 1: Directive*, version 6.0, Joint Service Publication 507, January 2014.

UK Ministry of Defence, *Financial Management and Charging Policy Manual, Part 1: Directive*, version 7.0, Joint Service Publication 462, March 2019a, withdrawn November 27, 2020.

UK Ministry of Defence, *Financial Management and Charging Policy Manual, Part 2: Guidance*, version 7.0, Joint Service Publication 462, March 2019b, withdrawn November 27, 2020.

UK Ministry of Defence, *How Defence Works*, version 6.0, September 2020a.

UK Ministry of Defence, *Multi-Domain Integration*, Joint Concept Note 1/20, November 2020b.

UK Ministry of Defence, *Defence in a Competitive Age*, March 2021a.

UK Ministry of Defence, *Defence and Security Industrial Strategy: A Strategic Approach to the UK's Defence and Security Industrial Sectors*, March 2021b.

UK Ministry of Defence, *Integrated Operating Concept*, August 2021c.

UK Ministry of Defence, *Annual Report and Accounts: 2020–21*, January 20, 2022a.

UK Ministry of Defence, *The Defence Equipment Plan: 2021–2031*, 2022b.

UK Ministry of Defence, *The Defence Equipment Plan: 2022 to 2032*, 2022c.

UK Ministry of Defence, "MOD Departmental Resources: 2021," webpage, February 24, 2022d. As of March 6, 2023:
https://www.gov.uk/government/statistics/defence-departmental-resources-2021/mod-departmental-resources-2021

UK Ministry of Defence, "The Defence and Security Public Contracts Regulations (DSPCR) 2011," webpage, updated November 28, 2022e. As of March 6, 2023:
https://www.gov.uk/government/publications/the-european-union-defence-and-security-public-contracts-regulations-dspcr-2011

UK Ministry of Defence, "DSPCR Chapter 9: Procuring Urgent Capability Requirements (UCRs)," webpage, updated September 27, 2023. As of October 15, 2023:
https://www.gov.uk/government/publications/the-european-union-defence-and-security-public-contracts-regulations-dspcr-2011/chapter-9-procuring-urgent-capability-requirements

UK Ministry of Defence, Defence Equipment and Support, "Who We Are," webpage, undated. As of March 6, 2023:
https://des.mod.uk/who-we-are

UK Ministry of Defence, Submarine Delivery Agency, "About Us," webpage, undated. As of March 6, 2023:
https://www.gov.uk/government/organisations/submarine-delivery-agency/about

UK National Audit Office, "Who We Are," webpage, undated. As of March 6, 2023:
https://www.nao.org.uk/about-us

UK National Audit Office, *Governance for Agile Delivery: Examples from the Private Sector*, July 2012.

UK National Audit Office, *Use of Agile in Large-Scale Digital Change Programmes: A Good Practice Guide for Audit and Risk Assurance Committees*, October 2022a.

UK National Audit Office, *The Equipment Plan 2022 to 2032: Ministry of Defence*, November 29, 2022b.

UK Parliament, "Erskine May," webpage, undated. As of March 6, 2023:
https://erskinemay.parliament.uk

UK Single Source Regulations Office, "About Us," webpage, undated. As of March 6, 2023:
https://www.gov.uk/government/organisations/single-source-regulations-office/about

UK Single Source Regulations Office, *Annual Compliance Report 2022*, November 2022.

U.S. Code, Title 10, Section 3131, Availability of Appropriations.

U.S. Department of Defense, *2022 National Defense Strategy of the United States of America*, 2022.

U.S. Department of State, "U.S. Relations with Canada," fact sheet, August 19, 2022.

U.S. House of Representatives, Committee on Appropriations, *Defense: Fiscal Year 2023 Appropriations Bill Summary*, U.S. Government Publishing Office, undated.

Warrell, Helen, "MoD Accused of Overspending as Budget 'Black Hole' Hits £17bn," *Financial Times*, January 12, 2021.

Watt, David, and Nicole Brangwin, "Defence Capability," webpage, Australian Parliament, undated. As of November 3, 2022:
https://www.aph.gov.au/About_Parliament/Parliamentary_Departments/Parliamentary_Library/pubs/BriefingBook46p/DefenceCapability

Webster, Adam, "Explainer: Can the Senate Block the Budget?" webpage, The Conversation, May 19, 2014. As of February 6, 2023:
https://theconversation.com/explainer-can-the-senate-block-the-budget-26815

Weight, Daniel, and Phillip Hawkins, *The Commonwealth Budget: A Quick Guide*, Australian Parliament, May 7, 2018.

Wezeman, Pieter D., Justine Gadon, and Siemon T. Wezeman, "Trends in International Arms Transfers, 2022," Stockholm International Peace Research Institute fact sheet, March 2023.

Wherry, Aaron, Rosemary Barton, David Cochrane, and Vassy Kapelos, "How the Liberals and New Democrats Made a Deal to Preserve the Minority Government," CBC News, March 27, 2022.

Wyatt, Austin, and Jai Galliott, *Toward a Trusted Autonomous Systems Offset Strategy: Examining the Options for Australia as a Middle Power,* Australian Army Occasional Paper No. 2, 2021.